Data in Science and Technology

Editor in Chief: R. Poerschke

Semiconductors

Group IV Elements and III-V Compounds

Editor: O. Madelung

Springer-Verlag Berlin Heidelberg New York
London Paris Tokyo
Hong Kong Barcelona

ISBN 3-540-53150-5 Springer-Verlag Berlin Heidelberg New York
ISBN 0-387-53150-5 Springer-Verlag New York Berlin Heidelberg

Library of Congress Cataloging-in-Publication Data
Semiconductors : group IV elements and III-V compounds / editor, O. Madelung.
p. cm. -- (Data in science and technology)
Includes bibliographical references.
ISBN 3-540-53150-5 (Springer-Verlag Berlin Heidelberg New York : acid-free paper)
ISBN 0-387-53150-5 (Springer-Verlag New York Berlin Heidelberg : acid-free paper)
1. Semiconductors--Handbooks, manuals, etc. I. Madelung, O. (Otfried) II. Series
QC611.45.S46 1991
537.6'22--dc20   90-26078 CIP

© Springer-Verlag Berlin Heidelberg 1991

Printed in Germany

Typesetting: Thomson Press, Ltd., New Delhi, India
Printing: Druckhaus Langenscheidt KG, Berlin
Bookbinding: Lüderitz & Bauer GmbH, Berlin

2163/3040-543210 – Printed on acid-free paper

Preface

The frequent use of wellknown critical data handbooks like Beilstein, Gmelin and Landolt-Börnstein is impeded by the fact that merely larger libraries – often far away from the scientist's working place – can afford such precious collections. To satisfy an urgent need of many scientists for having at their working place a comprehensive, high quality, but cheap collection of at least the basic data of their field of interest the series "Data in Science and Technology" is started now.

This first volume presents the most important data on two groups of semiconductors, the elements of the IVth group of the periodic system and the III-V compounds. All data were compiled from information on about 2500 pages in various volumes of the New Series of Landolt-Börnstein.

For each critically chosen data set and each figure the original literature is cited. In addition, tables of content refer to the handbooks the data were drawn from. Thus the presentation of data in this volume is of the same high quality standard as in the original evaluated data collections.

We hope to meet the needs of the physical community with the volumes of the series "Data in Science and Technology", forming bridges between the laboratory and additional information sources in the libraries.

Berlin/Marburg, February 1991   **The Editors**

Table of contents

A Introduction

1 General remarks

This volume contains basic data on Group IV and III–V Semiconductors. All data were compiled from volumes of the New Series of the Landolt-Börnstein data handbook. They comprise the information a scientist working on semiconductors is needing in his every-day work.

The data presented in this book are ordered under the following headings:

– Electronic properties:
  band structure / energies at symmetry points of the band structure / energy gaps (direct energy gap, indirect energy gap) / exciton energies / intra conduction band energies / intra valence band energies / critical point energies / spin-orbit splitting energies / camel's back structure of the conduction band edge / structure of the top of the valence band / effective masses (effective masses, electrons; effective masses, holes) / g-factor of electrons / valence band parameters.

– Lattice properties:
  structure / high pressure phases / transition temperature (pressure) or decomposition temperature / lattice parameters / linear thermal expansion coefficient / density / melting point / phonon dispersion relations / phonon frequencies (wavenumbers) / second order elastic moduli / third order elastic moduli.

– Transport properties:
  electrical conductivity (intrinsic conductivity) / (intrinsic) carrier concentration / carrier mobilities (electron mobility, hole mobility) / thermal conductivity (resistivity).

– Optical properties:
  optical constants / absorption coefficient / reflectance / extinction coefficient / refractive index / dielectric constants.

– Impurities and defects:
  solubility of impurities / diffusion coefficient (self-diffusion, impurity diffusion) / binding energies of (shallow) impurities / energy levels of impurities, defects and complexes or of deep centers.

  For alloys bowing parameters and crossover concentrations are also given.

Although the most relevant data have been summarized here the respective volumes of the New Series of Landolt-Börnstein contain much more information about these topics. In addition data on other properties can be found on the about 2500 pages of volumes 17a, 17c, 17d, 22a, 22b and 23a of Group III of the New Series:

Volume III/17a (and its supplement and extension **III/22a**) present data on group IV elements and III–V compounds. Additional information is given on topics as temperature and pressure dependence of energy band parameters, critical point energies, Kane and Luttinger band structure parameters, exciton parameters and deformation potentials; temperature and pressure dependence of lattice parameters, Debye temperatures, sound velocities, bulk modulus, Grüneisen parameters; carrier concentrations, drift velocities, galvanomagnetic, thermomagnetic and thermoelectric coefficients; optical constants and spectra, elasto- and piezooptic coefficients, Raman spectra; magnetic susceptibility, heat capacity, thermodynamical data and many other topics.

Volume III/22b is devoted to an extensive representation of all relevant data on impurities and defects in group IV and III–V semiconductors as solubilities and segregation constants, diffusion coefficients, shallow defect levels, deep defects and impurities, luminescence data, ESR and ENDOR data, local vibrational modes.

In addition to these physical data the **volumes III/17c and d** concentrate on technological data of the group IV and III–V semiconductors.

Chapter 2.1 of **volume III/23a** presents photoemission spectra and related band structure and core level data of tetrahedrally bonded semiconductors.

The organization and tables of contents of these volumes are described in the Appendix.

2 Physical quantities tabulated in this volume

Data on the following physical quantities are given in the tables and figures of Part B:

Sections on electronic properties

energies (unit eV):

$E(\boldsymbol{k})$ energy of a band state at wave vector $\boldsymbol{k}$.
Instead of the value of $\boldsymbol{k}$ often the respective symmetry point in the Brillouin zone is given (Γ, X, L, Σ...,
for the meaning of the symbols see Fig. 2 in section 1.1 for the diamond and zincblende structure,
Fig. 5 in section 2.1 for the wurtzite structure). Subscripts to these letters designate the irreducible
representation of the energy state (1, 1', 2, 12, 25'...). Indices c or v differentiate between states
lying in the conduction or valence band, respectively.

E_c, E_v energies of the edges of conduction and valence bands, respectively

E_g energy gap between conduction and valence band. Further subscripts refer to:

     dir: direct gap
     ind: indirect gap
     opt: optical gap (threshold energy for optical transitions)
     th: thermal gap (energy gap extrapolated to 0 K from transport measurements)
     x: excitonic gap (energy gap minus exciton binding energy)

E_b binding energy of the exciton

Δ mostly spin-orbit splittings of energy levels (subscripts 0, so, 1, 2 and dashes (') refer to the location
of the level as explained in the tables);
also other splittings of energy levels (cf: crystal-field splitting, ex: exciton exchange interaction energy,
LT: longitudinal-transverse exciton splitting energy)

$E_0...$ the letter E with other subscripts refers to intra- and interband transitions as explained in the tables
($E_0, E_1, E_2...$).

effective masses (in units of the electron mass m_0):

m_n, m_p effective mass of electrons (holes); other subscripts refer to:

     c: conductivity effective mass
     ds: density of states mass
     p,h: heavy holes
     p,l: light holes
     so: effective mass in the spin-orbit split band
     (X..):   effective mass at symmetry point X..

further conduction and valence band parameters:

ellipsoidal energy surfaces as occuring in the conduction band of group IV and III–V semiconductors are
characterized by the longitudinal and transverse effective masses

$m_\parallel, m_\perp$ defined by the equation $E(k) = E(k_0) + \hbar^2 \kappa_x^2/2m_\parallel + \hbar^2(\kappa_y^2 + \kappa_z^2)/2m_\perp$
where $\boldsymbol{\kappa} = \boldsymbol{k} - \boldsymbol{k}_0$ and $\kappa_x \parallel \boldsymbol{k}_0, \kappa_y, \kappa_z \perp \boldsymbol{k}_0$.

camel's back structure occurs at the conduction band edge in several III–V compounds. The relevant
parameters

$\Delta, \Delta_0, \Delta E, m_\parallel$ etc.   are explained in Fig. 2 and the accompanying equation in the tables of section 2.9.

warped energy surfaces as occuring in the valence band of group IV and III–V semiconductors are
characterized by valence band parameters

A, B, C defined by the equation $E(\boldsymbol{k}) = E(0) + (\hbar^2 k^2/2m_0)(A \pm (B^2 + sC^2)^{1/2})$
$$s = (k_x^2 k_y^2 + k_y^2 k_z^2 + k_z^2 k_x^2)/k^4.$$

g-factor of electrons: g_c

Sections on lattice properties

crystal lattice:

a, b, c lattice parameters (unit Å or nm)
α coefficient of linear thermal expansion (unit K^{-1})
d density (unit $g\,cm^{-3}$)
T_m melting temperature (unit K)
p_{tr} transition pressure for phase transitions (unit Pa)

phonon parameters:

ν phonon frequency (unit s^{-1})
$\bar{\nu}$ phonon wavenumber (unit cm^{-1})
$\nu(\boldsymbol{k})$ phonon dispersion relation (dependence of phonon frequency on wave vector); instead of $\boldsymbol{k}$ often the reduced wave vector $\zeta = k/k_{max}$ is used.
  Subscripts to the frequencies (wavenumbers) refer to transverse and longitudinal optical and acoustic branches (TO, LO, TA, LA) and to the symmetry points in the Brillouin zone as for the band structure energies.

elastic moduli:

c_{lm}, c_{lmn} second (third) order elastic moduli (unit $dyn\,cm^{-2}$)

Sections on transport properties

transport parameters:

R resistance (unit Ω)
R_H Hall coefficient (unit $cm^3\,C^{-1}$)
$\sigma, (\sigma_i)$ (intrinsic) electrical conductivity (unit $\Omega^{-1}\,cm^{-1}$)
ρ electrical resistivity (unit $\Omega\,cm$)
κ thermal conductivity (subscript L: lattice contribution) (unit $W\,cm^{-1}\,K^{-1}$)

carrier concentrations (unit cm^{-3}):

n electron concentration
p hole concentration
n_i intrinsic carrier concentration

carrier mobilities (unit cm^2/Vs):

μ_n, μ_p electron and hole mobilities, respectively. Further subscripts refer to:
  dr: drift mobility
  c: conductivity mobility
  H: Hall mobility

Sections on optical properties

optical constants:

K absorption coefficient (unit cm^{-1})
R reflectance (dimensionless)
n (real) refractive index (dimensionless)
k extinction coefficient (dimensionless)

ε dielectric constant; subscripts and brackets refer to:
   1:     real part of the complex dielectric constant
   2:     imaginary part of the complex dielectric constant
   (0):   low frequency limit
   (∞):   high frequency limit

Sections on impurities and defects

(Substitutional impurities are designated by (s), interstitial ones by (i))

solubility:

c^{eq} solubility of an impurity (maximum concentration incorporated in the solid in equilibrium without
          inducing a phase transition) (unit cm^{-3})
$c_0, \Delta H$ parameters of the Arrhenius equation $c^{eq} = c_0 \exp[-H/k_B T]$
          For a retrograde solubility a maximum solubility c^{eq}_{max} is observed at a temperature $T_{max} < T_m$.

diffusion coefficient:

D diffusion coefficient (unit $cm^2 s^{-1}$)
D_0, Q parameters occurring in the equation $D = D_0 \exp[Q/k_B T]$

energy levels:

E_b binding energy of donors $(E_c - E_d)$ or acceptors $(E_a - E_v)$
E for deep levels the type (d, a) is given; positive values refer to the valence band edge, negative values to
        the conduction band edge

B Physical data

1 Elements of the IVth group and IV–IV compounds

Physical property	Numerical value	Experimental conditions	Experimental method, remarks	Ref.

1.1 Diamond (C)

Electronic properties

band structure: Fig. 1 [84C] (Brillouin zone: Fig. 2)

Diamond is an indirect gap semiconductor, the lowest minima of the conduction band being located along the Δ-axes. The valence band has the structure common to all group IV semiconductors: three at Γ degenerate bands (spin neglected, symmetry $\Gamma_{25'}$). The spin orbit splitting of these bands is neglegible.

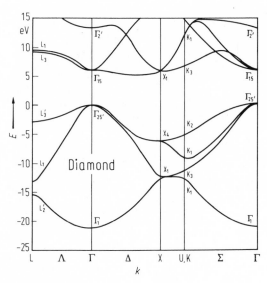

Fig. 1. Diamond. Band structure calculated by an ab intio LCAO method [84C].

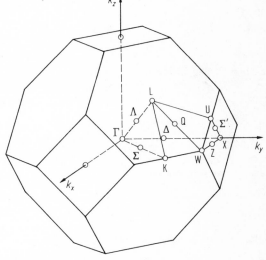

Fig. 2. Brillouin zone of the face centered cubic lattice, the Bravais lattice of the diamond and zincblende structures.

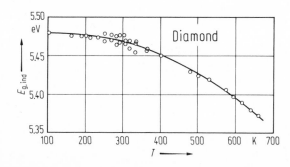

Fig. 3. Diamond. Indirect gap vs. temperature [64C].

Physical property	Numerical value	Experimental conditions	Experimental method, remarks	Ref.

energies of symmetry points of the band structure (relative to the top of the valence band) (in eV):

$E(\Gamma_{1v})$	$-21.03\ (-21(1))$		theoretical data:	
$E(\Gamma_{25'v})$	0.00		ab initio LCAO calculations	
$E(\Gamma_{15c})$	6.02 (6.0(2))		[84C], see Fig. 1	
$E(\Gamma_{2'c})$	13.41 (15.3(5))		experimental data	
$E(X_{1v})$	-12.43		(in brackets):	
$E(X_{4v})$	-6.27		photon energy dependent	
$E(X_{1c})$	5.91		photoemission [80H]	
$E(L_{1v})$	$-13.09\ (-12.8(3))$			
$E(L_{3'v})$	-2.82			
$E(L_{3c})$	9.23			
$E(L_{1c})$	9.58			

indirect energy gap:

$E_{g,ind}$	5.50(5) eV	RT	quantum photoyield	79H
$(\Gamma_{25'v} - \Delta_{1c})$				
$dE_{g,ind}/dT$	$-5 \cdot 10^{-5}$ eV/K	135···300 K	see Fig. 3	64C

Position of the minima of the conduction band along the Δ-axis: $k = (0.76(1), 0, 0)$ [65D].

direct energy gap:

$E_{g,dir}$	6.5 eV		angular dependent electron	82A
$(\Gamma_{25'v} - \Gamma_{15c})$			energy loss spectroscopy	
$dE_{g,dir}/dT$	$-6 \cdot 10^{-4}$ eV/K	135···300 K	absorption	64C

exciton binding energy:

E_b	0.080(5) eV		recombination radiation	65D

spin-orbit splitting energy:

$\Delta(\Gamma_{25'v})$	0.006(1) eV		cyclotron resonance	62R

effective masses, electrons (in units of m_0):

$m_{n\parallel}$	1.4		field dependence of	80N
$m_{n\perp}$	0.36		electron drift velocity	

effective masses, holes (in units of m_0):

$m_{p,h}$	1.08		calculated density of states	83R
$m_{p,1}$	0.36		from best set of valence	
m_{so}	0.15		band parameters	

valence band parameters:

$|A|$	3.61		most probable set of para-	83R
$|B|$	0.18		meters out of six sets	
$|C|$	3.67		published by various authors,	

Physical property	Numerical value	Experimental conditions	Experimental method, remarks	Ref.

Lattice properties

Structure

  The element carbon usually crystallizes in two modifications: diamond (cubic) and graphite (hexagonal) (Fig. 4). Under normal conditions (RT, atmospheric pressure) graphite is stable, diamond is metastable. According to their extrinsic properties diamonds are classified in several types (Ia, Ib, IIa, IIb).

High-pressure phases are not known.
Space group: O_h^7–Fd3m

lattice parameter:

a	3.56683(1) Å	298 K	X-ray diffraction	59K

For temperature dependence, see Fig. 5a.

linear thermal expansion coefficient: Fig. 5b.

density:

d	3.51525 g cm^{-3}		calculated from lattice constant	79F

melting point:

T_m	4100 K	$p = 12.5$ GPa	diamond-graphite-liquid eutectic	63B

phonon dispersion relations: Fig. 6.

phonon frequencies (in THz):

$\nu_{TO/LO}(\Gamma_{25'})$	39.9	300 K	Raman spectroscopy	70S
$\nu_{TA}(L_3)$	16.9			
$\nu_{LA}(L_1)$	30.2			
$\nu_{LO}(L_{2'})$	37.5			
$\nu_{TO}(L_{3'})$	36.2			
$\nu_{TA}(X_3)$	24.2			
$\nu_{LA/LO}(X_1)$	35.5			
$\nu_{TO}(X_4)$	32.0			

second order elastic moduli (in 10^{12} dyn cm^{-2}):

c_{11}	10.764(2)	296 K	Brillouin scattering	75G
c_{12}	1.252(23)			
c_{44}	5.774(14)			

Transport and optical properties

Most electrical and optical properties of diamond are *extrinsic*, i.e. strongly dependent on the impurity content. The most common impurity being nitrogen. Substitutional boron generates p-conductivity.

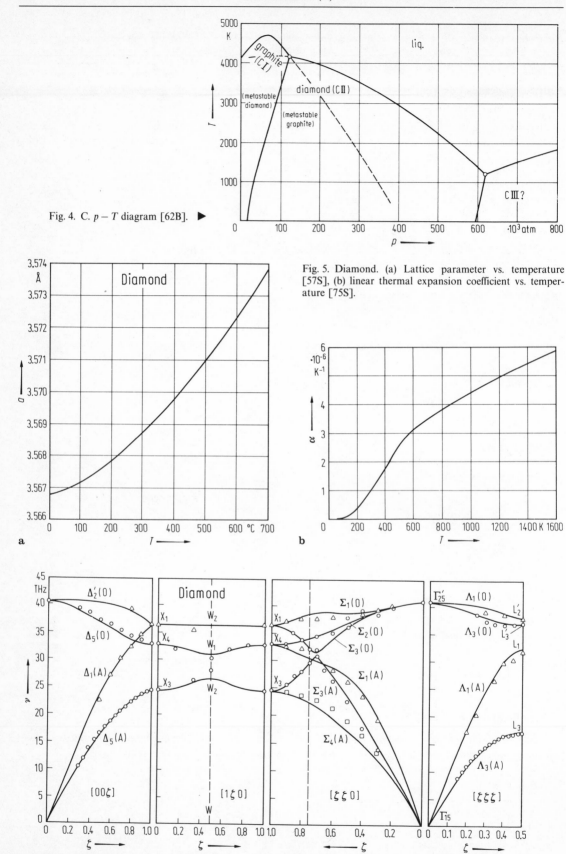

Fig. 4. C. $p - T$ diagram [62B]. ▶

Fig. 5. Diamond. (a) Lattice parameter vs. temperature [57S], (b) linear thermal expansion coefficient vs. temperature [75S].

Fig. 6. Diamond. Phonon dispersion relations. Experimental data from neutron scattering, full curves: shell model calculation [67W].

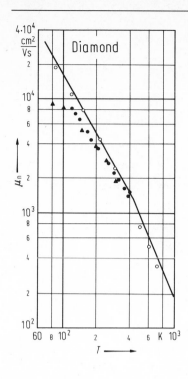

Fig. 7. Diamond. Electron mobility vs. temperature. Open circles: drift mobility data of [80N], full triangles and circles: Hall mobility data of [54R] and [67K], respectively. Continuous curve: theoretical drift mobility [80N].

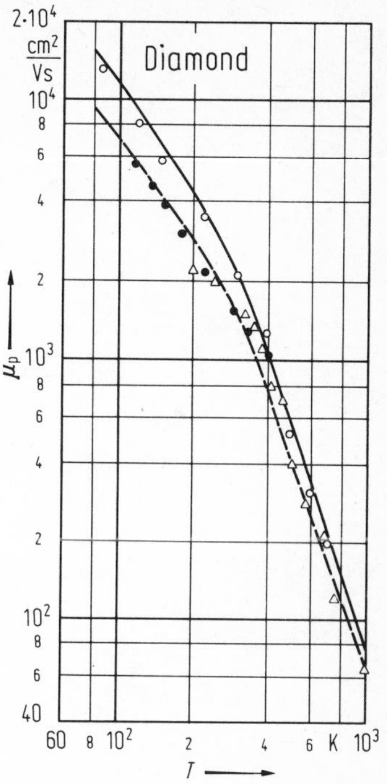

Fig. 8. Diamond. Hole mobility vs. temperature. Open circles: drift mobility data from [81R], full circles and triangles: Hall mobility data of [67K] and [65D], respectively. Solid and dashed curves: calculated drift and Hall mobilities, respectively [83R].

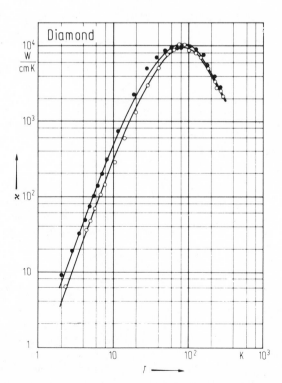

Fig. 9. Diamond. Thermal conductivity vs. temperature for two type IIa diamonds [76B].

Physical property	Numerical value	Experimental conditions	Experimental method, remarks	Ref.
electron mobility:				
μ_n	$\approx 2000 \, \mathrm{cm^2/Vs}$	RT	for temperature dependence of Hall and drift mobility, see Fig. 7	80N
hole mobility:				
μ_p	$2100 \, \mathrm{cm^2/Vs}$	RT	for temperature dependence of Hall and drift mobility, see Fig. 8	81R
	$\propto T^{-1.5}$	below 400 K		
	$\propto T^{-2.8}$	above 400 K		

thermal conductivity: mostly defect induced, see Fig. 9.

refractive index:

n	3.5	$\lambda = 177.0 \, \mathrm{nm}$		62P

n fits the empirical formula

$$n^2 - 1 = \frac{a\lambda^2}{\lambda^2 - \lambda_1^2} + \frac{b\lambda^2}{\lambda^2 - \lambda_2^2}: \quad a = 0.3306, \quad b = 4.3356, \quad \lambda_1 = 175.0 \, \mathrm{nm}, \quad \lambda_2 = 106.0 \, \mathrm{nm}$$

dielectric constant:

ε	5.70(5)	300 K	capacitance measurement at $10^3 \cdots 10^4$ Hz	77F

Temperature dependence: $\varepsilon = 5.70 - 5.35 \cdot 10^{-5} \, T + 1.66 \cdot 10^{-7} \, T^2$ (T in K) [77F].

Impurities and defects

Most electrical, optical and thermal properties of diamond are extrinsic, i.e. strongly dependent on the impurity content, the most important impurities being nitrogen and boron.

binding energies of impurities

Impurity	E_b[eV]	T[K]	Remarks	Ref.
Donors				
N	2	300	substitutional nitrogen; photoconductivity threshold	75V
	4 (1.45 eV above valence band)	300	nitrogen aggregates; photoconductivity threshold, pair recombination	67D 74C
Li	0.103(15)	300···600	ion implantation; conductivity measurements	79V
Acceptors				
B	0.370(10)	200···400	$n_a = 2 \cdot 10^{16} \, \mathrm{cm^{-3}}$; conductivity measurements	78M
	0.3685(15)	150···1250	$n_a \approx 5 \cdot 10^{16} \, \mathrm{cm^{-3}}$; Hall measurements	79C
	0.370	300···1000	synthetic diamond; $n_a = 2 \cdot 10^{16} \, \mathrm{cm^{-3}}$; conductivity measurements	79B

References for 1.1

54R   Redfield, A.G.: Phys. Rev. **94** (1954) 526.
57S   Skinner, B.J.: Am. Mineral. **42** (1957) 39.
59K   Kaiser, W., Bond, W.L.: Phys. Rev. **115** (1959) 857.
62B   Bundy, F.P.: Science **137** (1962) 1055.
62P   Philipp, H.R., Taft, E.A.: Phys. Rev. **127** (1962) 159.
62R   Rauch, C.J.: Proc. Int. Conf. Phys. Semicond., Exeter **1962** (A.C. Strickland ed.), The Inst. of Phys. and the Phys. Soc., London, p. 276.
63B   Bundy, F.P.: J. Chem. Phys. **38** (1963) 631.
64C   Clark, C.D., Dean, P.J., Harris, P.V.: Proc. Roy. Soc. London **A277** (1964) 312.
65D   Dean, P.J., Lightowlers, E.C., Wright, D.R.: Phys. Rev. **A140** (1965) 352.
67D   Denham, P., Lightowlers, E.C., Dean, P.J.: Phys. Rev. **161** (1967) 762.
67K   Konorova, E.A., Shevchenko, S.A.: Sov. Phys. Semicond. (English Transl.) **1** (1967) 299; Fiz. Tekh. Poluprovodn. **1** (1967) 364.
67W   Warren, J.L., Yarnell, J.L., Dolling, G., Cowley, R.A.: Phys. Rev. **158** (1967) 805.
70S   Solin, S.A., Ramdas, A.K.: Phys. Rev. **B1** (1970) 1687.
74C   Collins, A.T.: Ind. Diamond Rev. **34** (1974) 131.
75G   Grimsditch, M.H., Ramdas, A.K.: Phys. Rev. **B11** (1975) 3139.
75S   Slack, G.A., Bartram, S.F.: J. Appl. Phys. **46** (1975) 89.
75V   Vermeulen, L.A., Farrer, R.G.: Diamond Research 1975 (Suppl. to Ind. Diamond Rev.) 18.
76B   Berman, R., Martinez, M.: Diamond Research 1976 (Suppl. Ind. Diamond. Rev.) 7.
77F   Fontanella, J., Johnston, R.L., Colwell, J.H., Andeen, C.: Appl. Opt. **16** (1977) 2949.
78M   Massarani, B., Bourgoin, J.C., Chrenko, R.M.: Phys. Rev. **B17** (1978) 1764.
79F   Field, J.E. (ed.): The Properties of Diamond, Academic Press, London, New York, San Francisco **1979**.
79H   Himpsel, F.J., Knapp, J.A., van Vechten, J.A., Eastman, D.E.: Phys. Rev. **B20** (1979) 624.
79B   Bourgoin, J.C., Krynicki, J., Blanchard, B.: Phys. Status Solidi **(a)52** (1979) 293.
79C   Collins, A.T., Lightowlers, E.C.: The Properties of Diamond, Field, J.E. (ed.), London, New York, San Francisco: Academic Press **1979**, p. 79.
79V   Vavilov, V.S., Konorova, E.A., Stepanova, E.B., Trukhan, E.M.: Sov. Phys.-Semicond. **13** (1979) 635.
80H   Himpsel, F.J., van der Veen, J.F., Eastman, D.E.: Phys. Rev. **B22** (1980) 1967.
80N   Nava, F., Canali, C., Jacoboni, C., Reggiani, L.: Solid State Commun. **33** (1980) 475.
81R   Reggiani, L., Bosi, S., Canali, C., Nava, F.: Phys. Rev. **B23** (1981) 3050.
82A   Armon, H., Sellschop, J.P.F.: Phys. Rev. **B26** (1982) 3289.
83R   Reggiani, L., Waechter, D., Zukotynski, S.: Phys. Rev. **B28** (1983) 3550.
84C   Chelikowsky, J.R., Louie, S.G.: Phys. Rev. **B29** (1984) 3470.

Physical property	Numerical value	Experimental conditions	Experimental method, remarks	Ref.

1.2 Silicon (Si)

Electronic properties

band structure: Fig. 1 (Brillouin zone, see Fig. 2 of section 1.1)

The *conduction band* is characterized by six equivalent minima along the [100] axes of the Brillouin zone located at about $k_0 = 0.85\ (2\pi/a)$ (symmetry Δ_1). The surfaces of constant energy are ellipsoids of revolution with major axes along [100]. Higher minima are located at Γ and along the [111] axes about 1 eV above the [100] minima.

The *valence band* has its maximum at the Γ point (symmetry Γ_8), the (warped) light and heavy hole bands being degenerate at this point. The third spin-orbit split-off band has Γ_7-symmetry. The spin-orbit splitting energy at Γ is small compared to most interband energy differences. Thus, spin-orbit interaction is mostly neglected and the symmetry notation of the single group of the diamond lattice is used in the following tables.

energies of symmetry points of the band structure (relative to the top of the valence band) (in eV):

$E(\Gamma_{1v})$	$-12.34\ (-12.5(6))$	theoretical data:
$E(\Gamma_{25'v})$	0.00	localized atomic orbital
$E(\Gamma_{15c})$	3.50 (3.34···3.36)	calculation [85S], see Fig. 1
$E(\Gamma_{2'c})$	4.09 (4.15(5))	
$E(X_{1v})$	-7.75	

Physical property	Numerical value	Experimental conditions	Experimental method, remarks	Ref.
$E(X_{4v})$	−2.89 (−2.9)		experimental data (in brackets):	
$E(X_{1c})$	1.12 (1.13)		from a compilation in	
$E(L_{2'v})$	−9.62 (−9.3(4))		[83M], mostly photoemission	
$E(L_{1v})$	−7.01 (−6.8(2))		data	
$E(L_{3'v})$	−1.25 (−1.2(2))			
$E(L_{1c})$	2.29 (2.04(6))			
$E(L_{3c})$	4.34 (3.9(1))			

indirect energy gaps (in eV):

$E_{g,ind}$	1.1700	0 K (extrapol.)	wavelength modulated	74B
$(\Gamma_{25'v} - \Delta_{1c})$	1.1242	300 K	transmission	
$E_{g,th}$	1.205	0 K (extrapol.)	linear extrapolation from temperature dependence of conductivity above 200 K	67B

Temperature dependence $E_g(T) = A + BT + CT^2$ with $A = 1.170\,\text{eV}$, $B = 1.059 \cdot 10^{-5}\,\text{eV K}^{-1}$, $C = -6.05 \cdot 10^{-7}\,\text{eV K}^{-2}$ up to 300 K [74B]; see also Fig. 2 [85L].

$E_{g,ind}$	1.650(10) eV		optical absorption	74F
$(\Gamma_{25'v} - L_{1c})$	2.04(6) eV		cf. table above ("energies of symmetry points")	83M

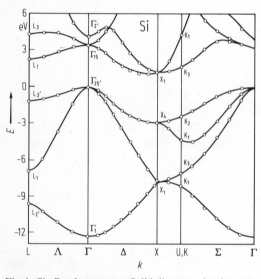

Fig. 1. Si. Band structure. Solid line: non-local, energy-dependent pseudopotential calculation, circles: localized atomic orbital method [85S].

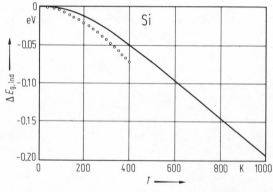

Fig. 2. Si. Temperature dependence of the indirect energy gap. Solid line: calculated, circles: experimental [85L]; $\Delta E_{g,ind} = E_{g,ind}(T) - E_{g,ind}(0)$.

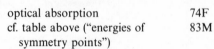

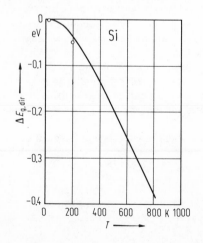

Fig. 3. Si. Temperature dependence of the direct band gap ($\Delta E_{g,dir} = E_{g,dir}(T) - E_{g,dir}(0)$). Solid line: calculated [85L], circles: experimental [72A].

Physical property	Numerical value	Experimental conditions	Experimental method, remarks	Ref.
exciton energies (in eV):				
$E(1S)$	1.15454(7)	1.6 K	photoluminescence emission	80S
E_b	0.0143(5)	1.8 K	exciton binding energy	77L2
direct energy gap (in eV):				
$E_{g,dir}$	4.185(10)	4.2 K	electroreflection	72A
$(\Gamma_{25'v} - \Gamma_{2'c})$	4.135	190 K		
$= E_0$	4.15···4.21		cf. table above ("energies of symmetry points")	83M

At higher temperatures E_0 cannot be optically resolved from the dominating E_2 critical point. Temperature dependence, see Fig. 3

spin-orbit splitting energy:				
$\Delta_0(\Gamma_{25'v})$	0.0441(3) eV	1.8 K	wavelength modulated absorption	74N
	0.045 eV	10 K	electroreflectance	78D
critical point energies (in eV):				
E'_0	3.378	10 K	ellipsometry	83J1

Direct transitions at Γ to higher conduction band. Nearly degenerate with the E_1 critical point, but different temperature shift. E'_0 and E_1 cannot clearly be resolved by experiment.

E_1	3.46(15)	10 K	electroreflectance	78D

Contributions mainly from transitions along the Λ-axes and at L.

E'_1	5.45	10 K	electroreflectance	78D

Transitions between $\Lambda_{3v} - \Lambda_{3c}$.

E_2	4.330	10 K	ellipsometry	83J

Contributions to E_2 from various transitions along the Σ and Δ axes.

effective masses, electrons (in units of m_0):				
$m_{n\parallel}$	0.1905(1)	1.26 K	cyclotron resonance, temperature dependence of transverse mass, see Fig. 4	65H
$m_{n\perp}$	0.9163(4)			
$m_{n,ds}$	1.026	4.2 K	cyclotron resonance, temperature dependence of density of states mass, see Fig. 5a	67B
g-factor of electrons:				
g_c	1.99893(28)		esr	66K
effective masses, holes (in units of m_0):				
$m_{p,h}$	0.537	4.2 K	cyclotron resonance	67B
$m_{p,l}$	0.153			
m_{so}	0.234			
$m_{p,ds}$	0.591	4.2 K	temperature dependence of density of states mass, see Fig. 5b	

In non-parabolic bands the density of states mass obtained through a calculation of the carrier concentration becomes temperature dependent, whereas the mass obtained directly from a calculation of the density of states

Fig. 4. Si. Transverse electron mass determined by cyclotron resonance vs. temperature [76O]. ▶

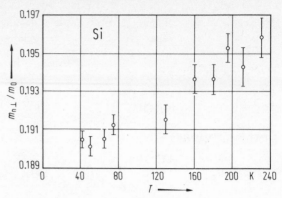

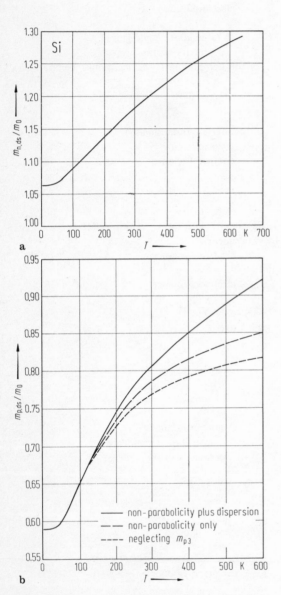

a

b

Fig. 5. Si. (a) Conduction band density of states mass vs. temperature for a sample with donor concentration below $5 \cdot 10^{17}$ cm^{-3}. (b) Valence band density of states mass vs. temperature as calculated from the experimentally determined valence band parameters [67B].

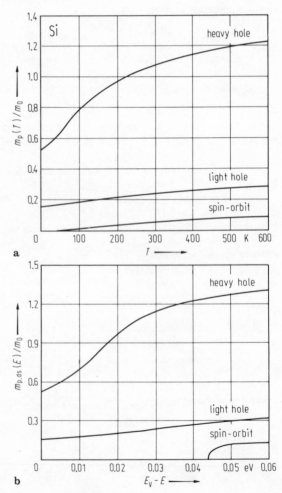

a

b

Fig. 6. Si. (a) Density of states mass obtained from carrier concentration vs. temperature for the three valence bands, (b) density of states mass obtained from the calculated density of states vs. energy measured from the top of the valence band. Both masses are related to each other by $m_p^{3/2}(T) = \langle m_{ds}^{3/2}(E) \rangle$, where $\langle \cdots \rangle$ means a weighted thermal average [83L].

Physical property	Numerical value	Experimental conditions	Experimental method, remarks	Ref.

becomes energy dependent. The former is the weighted thermal average of the latter: $m_p^{3/2}(T) = \langle m_{p,ds}^{3/2}(E) \rangle$. For the temperature and energy dependences of both masses, see Fig. 6.

valence band parameters:

A	$-4.27(2)$		cyclotron resonance	65B
B	$-0.63(8)$			
$|C|$	$4.93(15)$			

Lattice properties

Structure

Si I	space group O_h^7–Fd3m (diamond lattice)	stable at normal pressure	
Si II	space group D_{4h}^{19}–I4$_1$/amd	grey tin structure	
Si III	space group T_h^7–Ia3	metastable	84O1
Si IV	space group D_{6h}^4–P6$_3$/mmc		
Si V	space group D_{6h}^1–P6/mmm	simple hexagonal	84O1
Si VI		unidentified; stable above 34 GPa	84O1
Si VII		hexagonal close packed structure; stable above $\approx$ 45 GPa	84O1

transition pressures (in GPa):

p_{tr}	$11.2(2)\cdots12.5(2)$	Si I $\rightarrow$ Si II	84H,
	$13.2(2)\cdots16.4(5)$	Si II $\rightarrow$ Si V	85M1
	$14.5\cdots11.0$	Si V $\rightarrow$ Si II	
	$10.8\cdots8.5$	Si II $\rightarrow$ Si III	

lattice parameter (in nm):

a	$0.543102018(34)$	295.7 K	high purity single crystal measured in vacuum	82B1

temperature dependence of the lattice parameter: $a(T)$ in high-purity material can be approximated in the range $20\cdots800\,°C$ by $a(T) = 5.4304 + 1.8138\cdot10^{-5}(T - 298.15\,K) + 1.542\cdot10^{-9}(T - 298.15\,K)^2$ [73Y2], see also Fig. 7.

linear thermal expansion coefficient:

α	$2.59(5)\cdot10^{-6}\,K^{-1}$	298.2 K		84O2

Temperature dependence: Fig. 8 [84O2]. The data of Fig. 8 can be approximated in the temperature range $120\cdots1500\,K$ by the formula:
$\alpha(T) = (3.725\,[1 - \exp(-5.88\cdot10^{-3}\,[T - 124])] + 5.548\cdot10^{-4}\,T)\cdot10^{-6}\,K^{-1}$ (T in K).

density:

d	$2.329002\,g\,cm^{-3}$	25 °C	hydrostatic weighing, high purity crystal	64H

melting point:

T_m	$1685(2)\,K$		73H

phonon dispersion relations: Fig. 9.

Physical property	Numerical value	Experimental conditions	Experimental method, remarks	Ref.

phonon frequencies (in THz):

$\nu_{LTO}(\Gamma_{25'})$	15.53 (23)	296 K	from inelastic neutron	63D
$\nu_{TA}(X_3)$	4.49 (6)		scattering	
$\nu_{LAO}(X_1)$	12.32 (20)			
$\nu_{TO}(X_4)$	13.90 (30)			
$\nu_{TA}(L_3)$	3.43 (5)			
$\nu_{LA}(L_{2'})$	11.35 (30)			
$\nu_{LO}(L_1)$	12.60 (32)			
$\nu_{TO}(L_{3'})$	14.68 (30)			

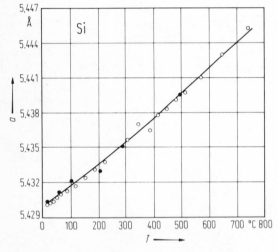

Fig. 7. Si. Lattice parameter vs. temperature in the range 20···740 °C, measurements on various samples [61H].

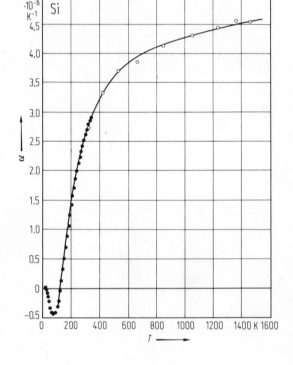

Fig. 8. Si. Linear thermal expansion coefficient vs. temperature. Experimental data from [77L1] (full circles) and [84O2] (open circles).

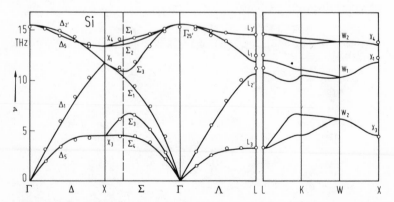

Fig. 9. Si. Phonon dispersion relation. Solid lines: theoretical [77W1], data points from [63D] and [72N].

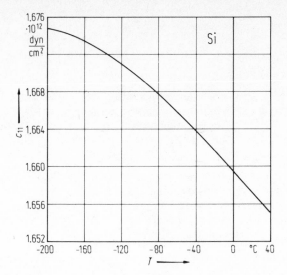

Fig. 10. Si. Second-order elastic modulus c_{11} vs. temperature [53M].

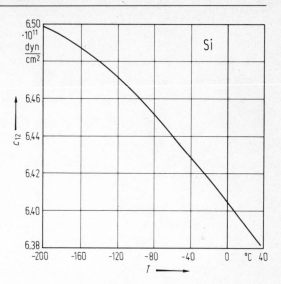

Fig. 11. Si. Second-order elastic modulus c_{12} vs. temperature [53M].

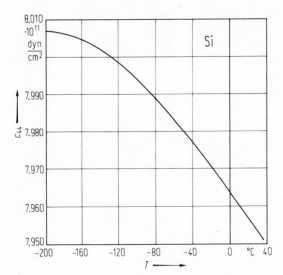

Fig. 12. Si. Second-order elastic modulus c_{44} vs. temperature [53M].

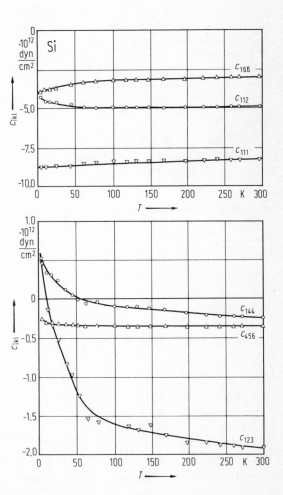

Fig. 13. Si. Third-order elastic moduli vs. temperature. Solid lines: best fit to the data [83P].

Physical property	Numerical value	Experimental conditions	Experimental method, remarks	Ref.

second order elastic moduli (in 10^{11} dyn cm^{-2}):

c_{11}	16.577	298 K	p-type sample, from ultrasound	64M
c_{12}	6.393	$\varrho = 410\,\Omega$ cm	sound measurements	
c_{44}	7.962		for temperature dependence,	
			see Figs. 10···12	

third order elastic moduli (in 10^{12} dyn cm^{-2}):

c_{111}	−8.34(11)	298 K	ultrasonic harmonic generation	81P
c_{112}	−5.31(32)		combined with pressure	
c_{144}	−0.95(24)		derivative of elastic constants	
c_{123}	−0.02(18)		for full temperature	
c_{166}	−2.96(12)		dependence, see Fig. 13	
c_{456}	−0.074(22)			

Transport properties

Contributions to the electric transport are exclusively made by electrons in the [100]-conduction band minima and (heavy and light) holes in the two uppermost valence bands. In samples with impurity concentrations below 10^{12} cm^{-3} the mobilities are determined by pure lattice scattering down to temperatures of about 10 K (n-type) or 50 K (p-type), respectively. For electrons the lattice mobility below 50 K is dominated by deformation potential coupling to acoustic phonons. At higher temperatures, intervalley scattering between the equivalent minima of the conduction band is added to the intravalley process, modifying the familiar $T^{-1.5}$ dependence to $T^{-2.42}$. At temperatures below 100 K, the lattice mobility of holes is dominated by acoustic scattering, but does not follow the $T^{-1.5}$ law due to the non-parabolicity of the valence bands. The proportionality of μ_p to $T^{-2.2}$ around RT is a consequence of optical phonon scattering.

intrinsic conductivity:

σ_i	$3.16\cdot10^{-6}\,\Omega^{-1}$ cm^{-1}	300 K		54M

The intrinsic conductivity up to 1273 K is given by the phenomenological expression $\log_{10} \sigma_i = 4.247 - 2.924 \cdot 10^3\,T^{-1}$ (σ_i in Ω^{-1} cm^{-1}, T in K).

intrinsic carrier concentration:

n_i	$1.02\cdot10^{10}$ cm^{-3}	300 K	see Fig. 14 for temperature dependence	77W2

n_i can be expressed in the range 200···500 K by $n_i(T) = 5.71\cdot10^{19}(T/300)^{2.365}\cdot\exp(-6733/T)$ cm^{-3} [77W2], in the range 450···700 K by $n_i(T) = 3.87\cdot10^{16}\,T^{3/2}\cdot\exp(-1.21/2\,kT)$ cm^{-3} (T in K, kT in eV).

electron mobility:

μ_n	1450 cm^2/Vs	300 K	lattice scattering mobility	54M

Around RT μ_n can be expressed by $\mu_n = 1.43\cdot10^9\,T^{-2.42}$ cm^2/Vs (T in K). For temperature dependence, see Fig. 15.

hole mobility (μ in cm^2/Vs):

$\mu_{H,p}$	370	300 K	ultrapure sample	82H,
	2.10^5	20 K	see Fig. 16 (b)	83S1,
				83S2,
$\mu_{c,p}$	505	300 K	see Fig. 16 (a)	83S3
	$1.6\cdot10^5$	20 K		

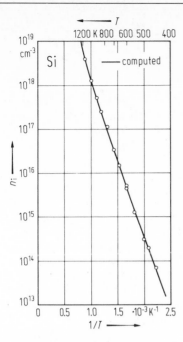

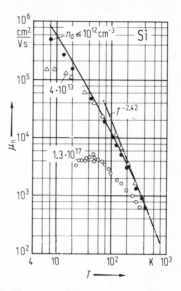

Fig. 14. Si. Carrier concentration in the intrinsic range vs. reciprocal temperature [54M].

Fig. 15. Si. Electron mobility vs. temperature; data points from three authors, solid line: theoretical lattice scattering mobility, dash-dotted line: $T^{-2.42}$ dependence of μ_n around RT [77J].

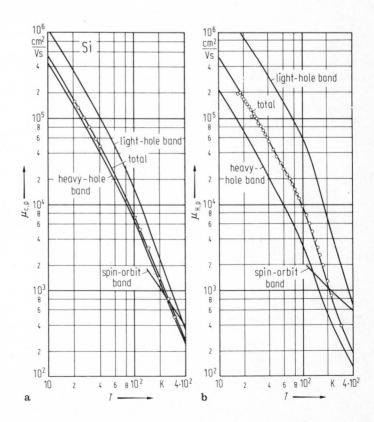

Fig. 16. Si. (a) Conductivity hole mobility vs. temperature (circles from [82M], triangles from [56L]). Solid lines: calculated contributions from the three valence bands. (b) Hall hole mobility vs. temperature (circles from [82M]). Solid lines as in (a) [83S1, 83S2].

Physical property	Numerical value	Experimental conditions	Experimental method, remarks	Ref.

thermal resistivity (in cm K W^{-1}):

κ_L^{-1}	$0.1598 + 1.532 \cdot 10^{-3}\,T$ $+ 1.583 \cdot 10^{-6}\,T^2$		$100\,\text{K} < T < 1200\,\text{K}$, lattice contribution	68F

Optical properties

dielectric constant:

		$T[\text{K}]$, $\nu[\text{MHz}]$		
ε	12.1	4.2, 750	capacitance bridge	66R
	11.9	300 750		

reflectance: in the temperature range $1100 \cdots 1400\,°\text{C}$ the reflectance can be approximated by $R = 30.37 + 1 \cdot 10^{-2}\,T - 1.3 \cdot 10^{-5}\,T^2 + 9.19 \cdot 10^{-10}\,T^3 + 5.65 \cdot 10^{-12}\,T^4$ (R in %, T in °C) [81L].
For the dependence of the absorption coefficient, extinction coefficient and refractive index on photon energy see Fig. 17.

refractive index: in the range $2.4 \cdots 25\,\mu\text{m}$ n can be approximated by the dispersion formula $n = A + BL + CL^2 + D\lambda^2 + E\lambda^4$ (λ in μm) with $L = (\lambda - 0.28)^{-1}\mu\text{m}^{-1}$, $A = 3.41983$, $B = 0.159906\,\mu\text{m}$, $C = -0.123109\,\mu\text{m}^2$, $D = 1.26878 \cdot 10^{-6}\,\mu\text{m}^{-2}$, $E = -1.95104 \cdot 10^{-9}\,\mu\text{m}^{-4}$ [80E].

optical constants:

real and imaginary parts of the dielectric constant measured by spectroscopical ellipsometry; n, k, R, K calculated from these data [83A]; see also Fig. 18.

$\hbar\omega[\text{eV}]$	ε_1	ε_2	n	k	R	$K[10^3\,\text{cm}^{-1}]$
1.5	13.488	0.038	3.673	0.005	0.327	0.78
2.0	15.254	0.172	3.906	0.022	0.351	4.47
2.5	18.661	0.630	4.320	0.073	0.390	18.48
3.0	27.197	2.807	5.222	0.269	0.461	81.73
3.5	22.394	33.818	5.610	3.014	0.575	1069.19
4.0	12.240	35.939	5.010	3.586	0.591	1454.11
4.5	−19.815	24.919	2.452	5.082	0.740	2317.99
5.0	−10.242	11.195	1.570	3.565	0.675	1806.67
5.5	−9.106	8.846	1.340	3.302	0.673	1840.59
6.0	−7.443	5.877	1.010	2.909	0.677	1769.27

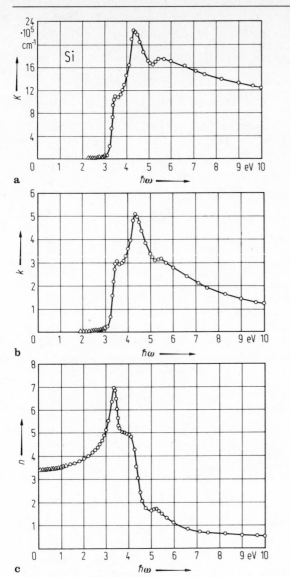

Fig. 17. Si. Optical constants below 10 eV vs. photon energy. (a) Absorption coefficient, (b) extinction coefficient, (c) refractive index [57C, 60P].

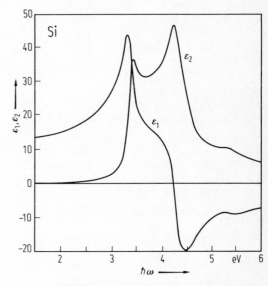

Fig. 18. Si. Real and imaginary parts of the dielectric constant vs. photon energy [83A].

Impurities and defects

The following tables on solubilities, diffusion coefficients and energy levels contain a representative selection of data and references from more than 50 pages on these topics in volume **NS III/22b**.

solubility of impurities in Si

Impurity	c_{max}^{eq} [cm^{-3}]	T_{max} [°C]	c_0 [cm^{-3}]	ΔH [eV]	T-Range [°C]	Ref.
Ag	10^{16}	1300	107	2.78	1024 ⋯ 1325	87R
Al	$2 \cdot 10^{19}$	1150				69W
As			$2.2 \cdot 10^{22}$	0.47	700 ⋯ 900	87A1
	$2.5 \cdot 10^{20}$	1000				83N
Au(s)	10^{17}	1300				88J
(i)	$1.5 \cdot 10^{15}$	1050	$8.2 \cdot 10^{24}$	2.56	700 ⋯ 1100	86G1
B			$3.2 \cdot 10^{22}$	0.43	870 ⋯ 1200	86D
	$8 \cdot 10^{20}$	1410				83W2
Bi	$8 \cdot 10^{17}$	1300				69W
C	$4 \cdot 10^{17}$	1400	$4 \cdot 10^{24}$	2.30	1000 ⋯ 1400	83K
Co	$3.5 \cdot 10^{6}$	1300	10^{26}	2.83	700 ⋯ 1250	83W1
Cr	10^{16}	1350	$5.5 \cdot 10^{24}$	2.79	800 ⋯ 1400	83W1
Cu	$1.5 \cdot 10^{18}$	1300	$5.5 \cdot 10^{23}$	1.49	500 ⋯ 1250	83W1
Fe	$2 \cdot 10^{16}$	1200	$1.8 \cdot 10^{16}$	2.94	900 ⋯ 1200	85W1
Ga	$6 \cdot 10^{20}$	1100				82B2
H	$7 \cdot 10^{15}$	1410				83W2
			$1.6 \cdot 10^{20}$	1.46	400 ⋯ 600	68I
In	$4 \cdot 10^{18}$	1100				82B2
Li	$6.5 \cdot 10^{19}$	1200				83W2
Mn	$1.5 \cdot 10^{16}$	1200	$5 \cdot 10^{25}$	2.78	900 ⋯ 1150	86G2
N	$4.5 \cdot 10^{15}$	1400				73Y1
Ni	$3 \cdot 10^{17}$	1100	$2.6 \cdot 10^{22}$	1.34	900 ⋯ 1200	86G1
O	$2 \cdot 10^{18}$	1400	$8.9 \cdot 10^{27}$	2.19	1350 ⋯ 1400	86C1
Os	$3.2 \cdot 10^{16}$	1260	$9 \cdot 10^{25}$	2.69	800 ⋯ 1000	87A2
P	$4 \cdot 10^{20}$	1100				83N
Pd	$4 \cdot 10^{15}$	1200	$1.8 \cdot 10^{20}$	1.34	900 ⋯ 1200	78S
Pt(s)	10^{17}	1250	$1.1 \cdot 10^{27}$	2.71	800 ⋯ 1000	75L
S	$2.5 \cdot 10^{17}$	1200				84W1
Sb	$5.3 \cdot 10^{19}$	1300	$3.81 \cdot 10^{21}$	0.56	850 ⋯ 1300	89N
Se	$7 \cdot 10^{16}$	1200	$5.5 \cdot 10^{27}$	2.84	800 ⋯ 1200	78V
Si	10^{18}	1400	10^{8}	4.40	700 ⋯ 1400	85T
Sn	$6 \cdot 10^{19}$	1100				69W
Ti	$2.3 \cdot 10^{13}$	1100	$6.6 \cdot 10^{23}$	2.85	1000 ⋯ 1250	86G1
Tl	$4 \cdot 10^{17}$	900				81K1
Vacancy	$6 \cdot 10^{17}$	1400	$8 \cdot 10^{23}$	2.0	800 ⋯ 1400	85T
Zn	$2.3 \cdot 10^{16}$	1100	$7.3 \cdot 10^{21}$	1.50	900 ⋯ 1360	86G1

(s): substitutional, (i): interstitial

diffusion coefficients of impurities in Si

Impurity	D_0 [$cm^2 s^{-1}$]	Q [eV]	T-Range [°C]	Ref.
Ag(i)	$6 \cdot 10^{-5}$	1.15	1024 ··· 1325	87R
Al	8.69	3.39	1060 ··· 1240	86S3
As	0.38	3.58	900 ··· 1400	77F2
Au(i)	$2.4 \cdot 10^{-4}$	0.39	300 ··· 1100	83G2
(s)	2.75	2.04	1100 ··· 1200	83G2
B	5.1	3.70	700 ··· 1200	83W1
Bi	3.96	4.12	1000 ··· 1200	77F1
C	1.9	3.10	1070 ··· 1400	83G1
Co	$9 \cdot 10^{-4}$	0.37	760 ··· 1100	88U
Cu(i)	$4.7 \cdot 10^{-3}$	0.43	400 ··· 900	86G1
(s)	$4 \cdot 10^{-3}$	1.0	800 ··· 1100	86G1
Fe	$1.3 \cdot 10^{-3}$	0.68	30 ··· 1300	83W1
Ga	6.3	3.48	1200 ··· 1250	85D
H	$4.2 \cdot 10^{-5}$	0.56	27 ··· 700	87P1
In	752	4.38	1000 ··· 1200	83C
K	$1.1 \cdot 10^{-8}$	0.80	500 ··· 800	72Z
Li	$2.5 \cdot 10^{-3}$	0.66	25 ··· 1350	83W1
N	0.87	3.29	1100 ··· 1200	76P
Na	0.01	1.27	500 ··· 850	76B
Ni(i)	$2 \cdot 10^{-3}$	0.47	800 ··· 1300	86G1
(s)	0.1	1.92	450 ··· 800	86G1
O	0.13	2.53	500 ··· 1400	86M2
P	10.5	3.69	1050 ··· 1350	83W1
S	0.22	2.10	1000 ··· 1230	88S2
Sb	30	4.08	1000 ··· 1200	87T
Se	0.3	2.60	1000 ··· 1310	88S1
Si	640	4.80	800 ··· 1200	86S1
Sn	0.054	3.50	1100 ··· 1200	74A
Te	0.9	3.30	1000 ··· 1310	88S1
Ti	0.015	1.8	950 ··· 1200	88M
Tl	351	4.26	1100 ··· 1338	77F1
Vac	0.6	4.03	700 ··· 1100	85T
Zn(i)	$3 \cdot 10^{-6}$	0.40	1100 ··· 1300	86G1

(s): substitutional, (i): interstitial

energy levels and structural information

on impurities, defects, and complexes in silicon. Negative energy refers to conduction band edge, positive energy refers to valence band edge.

Impurity, defect	E [eV]	Type	Remarks	Ref.
Al	+0.06903	a	T_d-symmetry, substitutional	83S4
	+0.17	d	interstitial Al after irradiation,	79T1
Al-C	+0.0563	a	aluminum-carbon pair, (X-center)	83S4
Al-Fe	+0.19	d	Fe(interstitial)-Al(substitutional) donor-acceptor pair, T_2-symmetry	88W
Al-H			passivation of Al substitutional acceptor	87S
As-Vac	−0.47	d	E-center, irradiation, anneals out at $T > 450$ K	79T2
Au	+0.346 +0.632 (−0.545)	d	substitutional	82W
B	+0.04439	a	substitutional acceptor	83F
B-C	+0.0371	a	boron-carbon pair, X-center	83S4
B-Fe	+0.100	d	B(substitutional)-Fe(interstitial)	86G1
	−0.23	a	donor-acceptor pair	
B-H			hydrogenation of B acceptor, atomic hydrogen	87S
Bi	−0.0710	d	substitutional, $1s(A_1)$ ground state,	85H1
Cd	+0.55	a	double substitutional acceptor Cd(0/−), single charge state	81D
	−0.45	a	double substitutional acceptor Cd(−/− −), double charge state	
Cr	−0.22	d	interstitial, Cr(0/+), $3d^6/3d^5$	86G1
	+0.128	d	double donor, interstitial?	81K2
Cu			many levels reported, uncertain identification	86G1
			inactive at $T < 850\,°C$, several irradiation defects at $T > 850\,°C$	86C2
			triple acceptor most likely	85F2
	+0.222		Cu-related, frequently reported values	86C2
	+0.411		passivated by hydrogen	85W2
Fe	+0.385	d	interstitial after quenching	86G1
Ga	+0.07273	a	substitutional acceptor	82S
Ga-C	+0.0572	a	Ga(substitutional)-C(substitutional) pair (X-center)	85F1
Ga-Fe	+0.14	d	Fe(interstitial)-Ga(substitutional)	86G1
	+0.23	d	donor-acceptor pair, configuration bistability	
Ga-H			passivation by atomic H	87S
In	+0.15558	a	substitutional acceptor,	83S4
In-C	+0.1128		In(substitutional)-C(substitutional), (X-center)	82S
In-Fe	+0.160	d	Fe(interstitial)-In(substitutional)	86G1
	+0.27	d	donor-acceptor pair	
In-H			passivation by atomic H,	87P1
Li	−0.03381	d	interstitial	81S
Li-Al			Li(interstitial)-Al(substitutional) donor-acceptor pair	65W1

(Continued)

Impurity, defect	E [eV]	Type	Remarks	Ref.
Li-B			Li(interstitial)-B(substitutional) donor-acceptor pair	65W2
Mg	-0.1075	d	interstitial Mg(0)	72H
	-0.2565	d	interstitial Mg($+$), T_d-symmetry	
Mn	$+0.26$	d	$Mn^{++/+}$ $3d^5$ interstitial,	86G1
	-0.42	d	$Mn^{+/0}$ $3d^6$	
	-0.12	a	$Mn^{0/-}$ $3d^7$	
	$+0.34$	d	$Mn^{+/0}$ $3d^2$ substitutional,	86G1
	-0.43	a	$Mn^{0/-}$ $3d^5$	
N			N-N pairs main incorporation	86S4
	-0.19	d	off-center substitutional partial	86S4
Ni			uncertain identification,	86G1
			15Ni-related complexes	87L
O			Si-O-Si bridging, displaced [111]-axis, electrically inactive	87P2
	-0.061	d	"thermal donor" TD, double donor $(0/+)$, formation at $350\cdots500\,°C$	87P2
	-0.132	d	"thermal donor" TD, double donor $(+/++)$, formation at $350\cdots500\,°C$, structure uncertain, C_{2v}-symmetry	87P2
O-Vac	-0.17	a	A-center, after irradiation C_{2v}-symmetry, (100)-orientation	87P2
P	-0.04558	d	substitutional, single donor	82G
	-0.0037	a	D^--center, binding of a second electron at 4 K	82N
P-Vac	-0.45	a	E-center, irradiation damage,	77K
Pd	-0.23	a	Pd($-/0$) predominant incorporation, $T=900\cdots1200\,°C$ annealing and quenching	86S2
	$+0.33$	d	Pd($+/0$) predominant incorporation, $T=900\cdots1200\,°C$ annealing and quenching	84L
Pd-H			passivation by atomic H	87P1
Pt	$+0.32\cdots+0.36$	d	substitutional Pt($+/0$)	87N
	$-0.23\cdots-0.26$	a	substitutional Pt($0/-$)	86S2
S	-0.3182	d	substitutional S($+/0$)double donor, T_d-symmetry	86P
	-0.6132	d	substitutional S($++/+$)double donor, T_d-symmetry	86P
S-S	-0.1875	d	sulfur pair $S_2(+/0)$ C_{3v}-symmetry	86P
	-0.3700	d	sulfur pair $S_2(++/0)$	86P
Sb-Vac	-0.44	a	E-center, irradiation, anneals at $T=460\,K$	79T2
Sb	-0.04277	d	substitutional single donor	84S
Se	-0.3065	d	substitutional double donor Se($+/0$), T_d-symmetry, 1s(A1)	86P
	-0.5932	d	substitutional double donor Se($++/+$), T_d-symmetry 1s(A1)	86P
Se-Se	-0.2064	d	Se-pair, double donor, $Se_2(+/0)$, D_{3D}-symmetry	86P
	-0.3892	d	Se-pair, double donor, $Se_2(++/+)$, D_{3D}-symmetry	86P

(*Continued*)

Impurity, defect	E [eV]	Type	Remarks	Ref.
Ta	-0.21			77B
Te	-0.1987	d	substitutional double donor $Te(+/0)$, T_d-symmetry	86P
	-0.4112	d	substitutional double donor $Te(++/+)$, T_d-symmetry	86P
Te-Te	-0.1580	d	double donor pair $Te_2(+/0)$, D_{3D}-symmetry	84W2
Ti	$+0.25 \cdots +0.28$	d	double donor $3d^3 - 3d^2$, interstitial	83W1
	-0.28	d	single donor $3d^4 - 3d^3$, interstitial	83W1
	-0.08	a	single acceptor $3d^5 - 3d^4$, interstitial	83W1
Tl	$+0.2460$	a	substitutional acceptor	83S4
Tl-C	$+0.1800$	a	X-center	83S4
Tl-H			passivation by atomic H	87P1
V	$+0.32$	d	double donor $V(++/+)$, interstitial $3d^4 - 3d^3$,	86G1
	-0.45	d	single donor $V(+/0)$, interstitial $3d^5 - 3d^4$,	86G1
Vac	$+0.05$	d	metastable center, single charge	84W3
Vac-Vac	$+0.20$	d	divacancy,	85H2
	-0.23	a	stable at $T \leqq 300\,°C$	
Zn	$+0.32$	a	double acceptor $Zn(0/-)$, substitutional	86G1
	-0.47 $(+0.66)$	a	double acceptor $Zn(-/--)$, substitutional	86G1

Reference for 1.2

53M   McSkimin, H.J.: J. Appl. Phys. **24** (1963) 988.
54M   Morin, F.J., Maita, J.P.: Phys. Rev. **96** (1954) 28.
56L   Ludwig, G.W., Watters, R.L.: Phys. Rev. **101** (1956) 1699.
57C   Carlson, R.O.: Phys. Rev. **108** (1957) 1390.
60P   Philipp, H.R., Taft, E.A.: Phys. Rev. **120** (1960) 37.
61H   Hall, R.O.A.: Acta Crystallogr. **14** (1961) 1004.
63D   Dolling, G.: in "Inelastic Scattering of Neutrons in Solids and Liquids", IAEA, Vienna **1963**, Vol. II, p. 37.
64H   Hennis, J.: J. Res. Nat. Bur. Stand. **68A** (1964) 529.
64M   McSkimin, H.J., Andreatch jr., P.: J. Appl. Phys. **35** (1964) 2161.
65B   Balslev, I., Lawaetz, P.: Phys. Lett. **19** (1965) 6.
65H   Hensel, J.C., Hasegawa, H., Nakayama, M.: Phys. Rev. **138** (1965) A225.
65W1   Weltzin, R.D., Swalin, R.A., Hutchinson, T.E.: Acta Metall. **13** (1965) 115.
65W2   Waldner, M., Hiller, M.A., Spitzer, W.G.: Phys. Rev. **A140** (1965) 172.
66K   Kodera, H.: J. Phys. Soc. Jpn. **21** Suppl. (1966) 578.
67B   Barber, H.D.: Solid State Electron. **10** (1967) 1039.
68F   Fulkerson, W., Moore, J.P., Williams, R.K., Graves, R.S., McElroy, D.L.: Phys. Rev. **167** (1968) 765.
68I   Ichimiya, T., Furuichi, T.: Int. J. Appl. Radiat. Isot. **19** (1968) 573.
69W   Wolf, H.F.: Semiconductors, New York: Wiley-Interscience **1971**.
72A   Aspnes, D.E., Studna, A.A.: Solid State Commun. **11** (1972) 1375.
72H   Ho, L.T., Ramdas, A.K.: Phys. Rev. **B5** (1972) 462.
72N   Nilsson, G., Nelin, G.: Phys. Rev. **B6** (1972) 3777.
72Z   Zorin, E.I., Pavlov, P.V., Tetelbaum, D.I., Khokhlov, A.F.: Fiz. Tekh. Poluprovodn. **6** (1972) 28.
73H   Hultgren, R., Desai, P.D., Hawkins, D. T., Gleiser, M., Kelly, K.K., Wagman, D.D.: The Thermodynamic Properties of the Elements, American Society for Metals, Metals Park, Ohio **1973**.
73Y1   Yatsurugi, Y., Akiyama, N., Endo, Y., Nozaki, T.: J. Electrochem. Soc. **120** (1973) 975.
73Y2   Yin, W.M., Paff, R.J.: J. Appl. Phys. **45** (1973) 1456.
74A   Akasaka, Y., Horie, K., Nakamura, G.: Jpn. J. Appl. Phys. **13** (1974) 1533.
74B   Bludau, W., Onton, A., Heinke, W.: J. Appl. Phys. **45** (1974) 1846.
74F   Foreman, R.A., Aspnes, D.E.: Solid State Commun. **14** (1974) 100.
75L   Lisiak, K.P., Milnes, A.G.: Solid State Electron. **18** (1975) 533.
74N   Nishino, T., Takeda, M., Hamakawa, Y.: Solid State Commun. **14** (1974) 627.

76B   Belikova, M.N., Zastavnyi, A.V., Korol, V.M.: Fiz. Tekh. Poluprovodn. **10** (1976) 535.
76H   Hensel, J.C.: unpublished.
76O   Ousset, J.C., Leotin, J., Askenasy, S., Skolnick, M.S., Stradling, R.A.: J. Phys. **C9** (1976) 2802.
76P   Pavlov, P.V., Zorin, E.I., Tetelbaum, D.I., Khokhlov, A.F.: Phys. Status Solidi (a) **35** (1976) 11.
77B   Busta, H.H., Waggener, H.A.: J. Electrochem. Soc. **124** (1977) 1424.
77F1   Fair, R.B.: Semiconductor Silicon 1977, Huff. H.R., Sirtl, E. (eds.), The Electrochem Soc. **1977** p. 968.
77F2   Fair, R.B., Tsai, J.C.C.: J. Electrochem. Soc. **124** (1977) 1107.
77J   Jacoboni, C., Canali, C., Ottaviani, G., Alberigi Quaranta, A.: Solid State Electron. **20** (1977) 77.
77K   Kimerling, L.C.: Radiation Effects in Semiconductors 1976, in: Inst. Phys. Conf. Ser. **31** (1977) 221.
77L1   Lyon, K.G., Salinger, G.L., Swenson, C.A., White, G.K.: J. Appl. Phys. **48** (1977) 865.
77L2   Lipari, N.O., Altarelli, M.: Phys. Rev. **B15** (1977) 4883.
77W1   Weber, W.: Phys. Rev. **B15** (1977) 4793.
77W2   Wasserrab, Th.: Z. Naturforsch. **32a** (1977) 746.
78D   Daunois, A., Aspnes, D.E.: Phys. Rev. **B18** (1978) 1824.
78S   So, L., Whiteley, J.S., Ghandi, S.K., Baliga, B.J.: Solid State Electron. **21** (1978) 887.
78V   Vydianath, H.R., Lorenzo, J.S., Kröger, F.A.: J. Appl. Phys. **49** (1978) 5928.
79T1   Troxel, J.R., Chatterjee, A.P., Watkins, G.D.: Phys. Rev. **B19** (1979) 5336.
79T2   Troxel, J.R.: Ph.D. Thesis, Lehigh University, U.S.A. **1979**.
80E   Edwards, D.F., Ochoa, E.: Appl. Opt. **19** (1980) 4130.
80S   Schmid, W.: Phys. Rev. Lett. **45** (1980) 1726.
81D   Dyunaidov, S.S., Urmanov, N.A., Gafurova, M.V.: Phys. Status Solidi (a) **66** (1981) K79.
81K1   Keller, W.: Diplomarbeit Univ. Erlangen **1981**.
81K2   Kunio, T., Nishino, T., Ohta, E., Sakata, M.: Solid State Electron. **24** (1981) 1087.
81L   Lampert, M.O., Koebel, J.M., Siffert, P.: J. Appl. Phys. **52** (1981) 4975.
81P   Philip, J., Breazeale, M.A.: J. Appl. Phys. **52** (1981) 3383.
81S   Szablak, B., Altarelli, M.: Solid State Commun. **37** (1981) 341.
82B1   Becker, P., Seyfried, P., Siegert, H.: Z. Physik **B48** (1982) 17.
82B2   Budzak, Ya.S., Mavrin, O.I.: Phys. Status Solidi (a) **69** (1982) K61.
82G   Grimmeiss, H.G., Janzén, E., Larsson, K.: Phys. Rev. **B25** (1982) 2627.
82H   Haug, A., Schmid, W.: Solid State Electron. **25** (1982) 665.
82M   Mitchel, W.C., Hemenger, P.M.: J. Appl. Phys. **53** (1982) 6880.
82N   Narita, S., Shinbashi, T., Kobayashi, M.: J. Phys. Soc. Jpn. **51** (1982) 2186.
82S   Searle, C.W., Ohmer, M.C., Hemenger, P.M.: Solid State Commun. **44** (1982) 1597.
82W   Wu, R.H., Peaker, A.R.: Solid State Electron. **25** (1982) 463.
83A   Aspnes, D.E., Studna, A.A.: Phys. Rev. **B27** (1983) 985.
83C   Cerofolini, G.F., Pignatel, G.U., Riva, F.: Thin Solid Films **10** (1983) 275.
83F   Fischer, D.W., Rome, J.J.: Phys. Rev. **B27** (1983) 4826.
83G1   Gösele, U., Tan, T.Y.: Aggregation Phenomena of Point Defects in Si, Sirtl, E. (ed.), The Electrochem. Soc. **1983**, p. 17.
83G2   Graff, K.: Aggregation Phenomena of Point Defects in Si, Sirtl, E. (ed.), The Electrochem. Soc. **1983**, p. 121.
83J   Jellison, G.E., Modine, F.A.: Phys. Rev. **B27** (1983) 7466.
83K   Kolbesen, B.O.: Aggregation Phenomena of Point Defects in Si, Sirtl, E. (ed.), The Electrochem. Soc. **1983**, p. 155.
83L   Lang, J.E., Madarasz, F.L., Hemenger, P.M.: J. Appl. Phys. **54** (1983) 3612.
83M   Mašović, D.R., Vukajlović, F.R., Zeković, S.: J. Phys. **C16** (1983) 6731.
83N   Nobili, D.: Aggregation Phenomena of Point Defects in Si, Sirtl, E. (ed.), The Electrochem Soc. **1983**, p. 189.
83P   Philip, J., Breazeale, M.A.: J. Appl. Phys. **54** (1983) 752.
83S1   Szmulowicz, F.: Appl. Phys. Lett. **43** (1983) 485.
83S2   Szmulowicz, F.: Phys. Rev. **B28** (1983) 5943.
83S3   Szmulowicz, F., Madarasz, F.L.: Phys. Rev. **B27** (1983) 2605.
83S4   Searle, C.W., Hemenger, P.M., Ohmer, M.C.: Solid State Commun. **48** (1983) 995.
83W1   Weber, E.: Appl. Phys. **A30** (1983) 1.
83W2   Wacker Chemitronics Co.: Silicon calculator **1983**.
84H   Hu, J.Z., Spain, I.L.: Solid State Commun. **44** (1984) 263.
84L   Lemke, H.: Phys. Status Solidi (a) **86** (1984) K39.
84O1   Olijnuk, H., Sikka, S.K., Holzapfel, W.B.: Phys. Lett. **A103** (1984) 137.
84O2   Okada, Y., Tokumaru, Y.: J. Appl. Phys. **56** (1984) 314.
84S   Scalar, N.: Appl. Phys. **55** (1984) 2972.
84W1   Wagner, P., Holm, C., Sirtl, E., Oeder, R., Zulehner, W.: Festkörperprobleme **XXIV** (1984) 191.
84W2   Wagner, P., Holm, C.: 13th Int. Conf. on Defects in Semiconductors **1984**.
84W3   Watkins, G.D.: Festkörperprobleme XXIV, Grosse, P. (ed.), Braunschweig: Vieweg **1984**, p. 163.
85D   Danilicheva, T.A., Markvicheva, V.S., Nisnevich, J.D.: Izv. Akad. Nauk SSSR Neorg. Mater. **21** (1985) 525.
85F1   Fischer, D.W., Mitchel, W.C.: J. Appl. Phys. **58** (1985) 3118.
85F2   Fazzio, A., Caldas, M.J., Zunger, A.: Phys. Rev. **B32** (1985) 934.
85H1   Hertel, N., Materlik, G., Zegenhagen, J.: Z. Phys. **B58** (1985) 199.
85H2   Harris, R.D., Watkins, G.D.: Proc. Defect. Conf. Coronado, Kimmerling, L.C., (ed.), The Met. Soc. of AIME **1985**, p. 799.
85L   Lautenschlager, P., Allen, P.B., Cardona, M.: Phys. Rev. **B31** (1985) 2163.
85S   Sieh, K.S., Smith, P.V.: Phys. Status Solidi (b) **129** (1985) 259.
85T   Tan, T.Y., Gösele, U.: Appl. Phys. **A37** (1985) 1.
85W1   Weber, E.R.: Proc. SPIE (Proc. Soc. Photo-Opt. Instrum. Eng.) **524** (1985) 160.
85W2   Wang, Z., Chen, K., Qin, G.: Chin. J. Semicond. **6** (1985) 437.

86C1	Carlberg, T.: J. Electrochem. Soc. **133** (1986) 1941.
86C2	Chen, K.-M., Qin, G.-G.: Proc. 14th Int. Conf. on Defects in Semiconductors, Paris **1986**.
86D	Dominguez, E., Jaraiz, M.: J. Electrochem. Soc. **133** (1986) 1895.
86G1	Graff, K.: Semiconductor Silicon 1986, H.R. Huff et al. (eds.), The Electrochem. Soc. **1986**, p. 751.
86G2	Gilles, D., Bergholz, W., Schröter, W.: J. Appl. Phys. **59** (1986) 3590.
86M1	Mononi, C.S., Hu, J.Z., Spain, I.L.: Phys. Rev. **B34** (1986) 362.
86M2	Mikkelsen jr., J.C.: Mater. Res. Soc. Symp. Proc. **59** (1986) 19.
86P	Pensl, G., Roos, G., Holm, C., Wagner, P.: Proc. 14th Int. Conf. on Defects in Semiconductors, Paris **1986**.
86S1	Stolwijk, N.A., Hölzl, J., Frank, W., Weber, E.R., Mchrer, H.: Appl. Phys **A39** (1986) 37.
86S2	Stöffler, W., Weber, J.: Proc 14th Int. Conf. on Defects in Semiconductors, Paris **1986**.
86S3	Schulz, H.J.: European Semiconductor Device Research Conference (ESSDERC) **1986**.
86S4	Stein, H.J.: Proc. MRS Meeting Boston 1986, Mikkelsen, J. (ed.) MRS Pittsburgh, Pa. **1986**, p. 523.
87A1	Angelucci, R., Armigliato, A., Landi, E., Nobili, D., Solmi, S.: ESSDERC Conference Bologna **1987**.
87A2	Azomov, S.A., Yunusov, M.S., Nurkuziev, G.: Fiz. Tekh. Poluprovodn. **21** (1987) 1555; Sov. Phys. Semicond. (English Transl.) **21** (1987) 944.
87L	Lemke, H.: Phys. Status Solidi (a) **99** (1987) 205.
87N	Nolte, D.D., Walukiewicz, W., Haller, E.E.: Phys. Rev. **B36** (1987) 9392.
87P1	Pearton, S.J., Corbett, J.W., Shi, T.S.: Appl. Phys. **A43** (1987) 153.
87P2	Pensl, G.: Proc. 5th Int. School ISPPME **1987**.
87R	Rollert, F., Stolwijk, N.A., Mehrer, H.: J. Phys. **D20** (1987) 1148.
87S	Stavola, M., Pearton, S.J., Lopata, J., Dautremont-Smith, W.C.: Appl. Phys. Lett. **50** (1987) 1086.
87T	Tsai, J.C.C., Schimmel, D.G., Fair, R.B., Maszara, W.: J. Electrochem. Soc. **134** (1987) 1508.
88J	Jäntsch, O.: private communication.
88M	Mathiot, D., Hocine, S.: 15. Int. Conf. Defects in Semiconductors Budapest 1988.
88S1	Stuempel, H., Vorderwuelbecke, M., Mimkes, J.: Appl. Phys. **A46** (1988) 159.
88S2	Stolwijk, N.A., Grünebaum, D., Perret, M., Brohl, M.: Proc. 15th Int. Conf. on Defects in Semiconductors, Budapest 1988, Trans. Tech. Publ. **1988**.
88U	Utzig, J., Gilles, D.: Proc. 15th Int. Conf. on Defects in Semiconductors, Budapest 1988, Ferency, G. (ed.), Trans. Tech. Publ. **1988**.
88W	Watkins, G.D.: Proc. 15th Int. Conf. on Defects in Semiconductors, Budapest 1988, Ferency, G. (ed.), Trans. Tech. Publ. **1988**.
89N	Nobili, D., Angelucci, R., Armigliato, A., Landi, E., Solmi, S.: J. Electrochem. Soc. **136** (1989) 1142.

Physical property	Numerical value	Experimental conditions	Experimental method, remarks	Ref.

1.3 Germanium (Ge)

Electronic properties

band structure: Fig. 1 (Brillouin zone, see Fig. 2 of section 1.1).

The *conduction band* is characterized by eight equivalent minima at the end points of the [111]-axes of the Brillouin zone (symmetry L_6). The surfaces of constant energy are ellipsoids of revolution with major axes along [111]. Higher minima are located at the Γ-point and (above this) on the [100]-axes.

The *valence band* has its maximum at the Γ-point (symmetry Γ_8), the (warped) light and heavy hole bands being degenerate at this point. The third spin-orbit split-off band has Γ_7-symmetry. The spin-orbit splitting is considerable. Thus, the symmetry notation of the double group of the diamond lattice is used in the following tables.

energies of symmetry points of the band structure (relative to the top of the valence band) (in eV):

$E(\Gamma_{6v})$	-12.66	theoretical data (Fig. 1)	76C
$E(\Gamma_{7v})$	-0.29		
$E(\Gamma_{8v})$	0.00		
$E(\Gamma_{7c})$	0.90		
$E(\Gamma_{6c})$	3.01		
$E(\Gamma_{8c})$	3.22		
$E(X_{5v})$	-8.65	for experimental data from	
$E(X_{5v})$	-3.29	angular resolved photo-	
$E(X_{5c})$	1.16	emission, see [85W2]	
$E(L_{6v})$	-10.39	and [84H, 85N]	

Physical property	Numerical value	Experimental conditions	Experimental method, remarks	Ref.
$E(L_{6v})$	−7.61			
$E(L_{6v})$	1.63			
$E(L_{4,5v})$	−1.43			
$E(L_{6c})$	0.76			
$E(L_{6c})$	4.16			
$E(L_{4,5c})$	4.25			

indirect energy gap (in eV):

$E_{g,ind}(\Gamma_{8v} - L_{6c})$	0.744 (1)	1.5 K	magnetotransmission	59Z
	0.664	291 K	optical absorption	57M
$E_{g,th}$	0.785	0 K (extrapol.)	temperature dependence of intrinsic conductivity	54M1

Temperature dependence, see Fig. 2.

direct energy gap:

$E_{g,dir}(\Gamma_{8v} - \Gamma_{7c})$	0.898 (1) eV	1.5 K	magnetoabsorption	59Z
	0.805 (1) eV	293 K		

Temperature dependence, see Fig. 3.

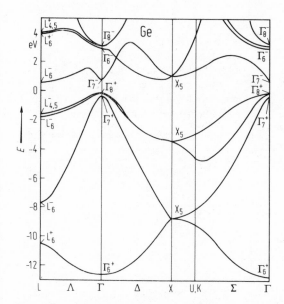

Fig. 1. Ge. Band structure obtained by a non-local pseudo-potential calculation including spin-orbit splitting [76C].

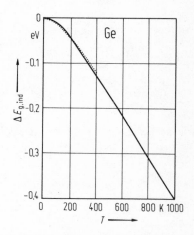

Fig. 2. Ge. Indirect gap vs. temperature. Solid line: calculated, dots: experimental $(\Delta E_{g,ind} = E_{g,ind}(T) - E_{g,ind}(0))$ [85L].

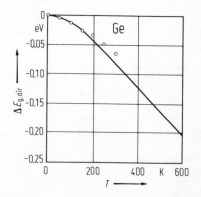

Fig. 3. Ge. Direct gap vs. temperature $(\Delta E_{g,dir} = E_g(T) - E_g(0))$. Solid line: theoretical calculations, circles: experimental [83A2].

Physical property	Numerical value	Experimental conditions	Experimental method, remarks	Ref.

exciton energies (in meV):

$E(1S^{3/2}_{3/2}(L_4 + L_5))$	740.46(3)	2.1 K, 5.1 K	absorption (at 2.1 K) and	79M
$E(1S^{3/2}_{1/2}(L_6))$	741.58(3)		luminescence (at 5.1 K), the energy of the assisting LA (L) phonon is subtracted	
E_b	4.18		$1S^{3/2}_{3/2}(L_4 + L_5)$	76A
	3.17		$1S^{3/2}_{1/2}(L_6)$	

critical point and spin-orbit splitting energies (in eV):

E_1	2.111(3)	RT	from ellipsometric data by a two-dimensional critical point analysis (2D CP)	84V
$E_1 + \Delta_1$	2.298(3)	RT	2D CP	
Δ_1	0.187(3)	RT	2D CP	
E'_0	3.123(19)	100 K	3D CP	
$E'_0 + \Delta'_0$	3.309(19)	100 K	3D CP	
$\bar{E}'_0$	3.110	RT	2D CP; mean value of E'_0 and $E'_0 + \Delta'_0$	
E_2	4.368(4)	RT	2D CP	
	4.346(3)	RT	1D CP	

effective masses, electrons (in units of m_0):

$m_{n\perp}(L_6)$	0.0807(8)	30⋯100 K	cyclotron resonance	76F
	0.0823	120 K	magnetophonon resonance	82H2
$m_{n\parallel}(L_6)$	1.57(3)	30⋯100 K	cyclotron resonance	76F
	1.59	120 K	magnetophonon resonance	82H2
$m_n(\Gamma_7)$	0.0380(5)	30 K	piezomagnetoreflectance	70A

For the dependence of the transverse electron mass in the L_6-minima on the energy above the bottom of the band, see Fig. 4a.

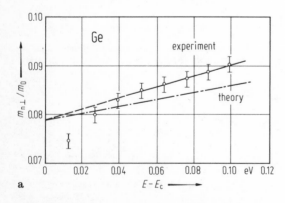

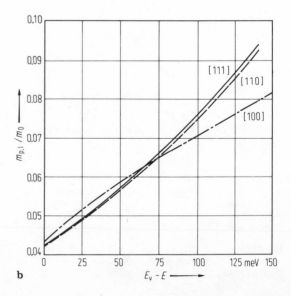

Fig. 4. Ge. (a) Dependence of the transverse electron mass in the L_6 minima on energy above the bottom of the conduction band measured by magnetopiezotransmission [69A], (b) cyclotron mass of light holes vs. energy below the top of the valence band ($B \parallel$ [111], [110] and [100]) [80Z].

Physical property	Numerical value	Experimental conditions	Experimental method, remarks	Ref.
g-factor of electrons:				
g_c	$-3.0(2)$	30 K	piezomagnetoabsorption	70A
effective masses holes (in units of m_0):				
$m_{p,1}$	0.0438(3) 0.0426(2) 0.0430(3)	4 K	cyclotron resonance	56D
$m_{p,h}$	0.284(1) 0.376(1) 0.352(4)		values for B parallel [100], [111], [110], respectively	
m_{so}	0.095(7)	30 K	piezomagnetoabsorption	70A

For the dependence of the light hole mass on the energy below the top of the band, see Fig. 4b.

valence band parameters:				
A	-13.3		extrapolated using a five level $k \cdot p$ scheme	75W
B	-8.57			
$|C|$	12.78			

Lattice properties

Structure

Ge I	space group O_h^7–Fd3m (diamond lattice)		low pressure phase	
G II	space group D_{4h}^{19}–I4$_1$/amd (β-tin lattice)		from energy dispersive X-ray diffraction	83Q
Ge III	space group D_4^8–P4$_3$2$_1$2		obtained by quenching of Ge II to lower pressure	83Q
Ge IV	space group T_h^7–Ia3 (body centered cubic)		obtained by quenching of Ge II to 1 atm at 200 K	83L

transition pressure:				
p_{tr}(Ge I–Ge II)	10.7(5) GPa		diamond anvil, energy dispersive X-ray diffraction	82W

lattice parameter:				
a	5.6579060 Å	298.15 K	single crystal, for temperature dependence see Fig. 5	75B

linear thermal expansion coefficient: Fig. 6.

density:				
d	5.3234 g cm^{-3}	298 K	hydrostatic weighing	52S

melting point:				
T_m	1210.4 K			73H

phonon dispersion relations: Fig. 7.

Physical property	Numerical value	Experimental conditions	Experimental method, remarks	Ref.
phonon frequencies (in THz):				
$v_{LTO}(\Gamma_{25'})$	9.02(2)	300 K	coherent inelastic neutron scattering	72N2
$v_{TA}(X_3)$	2.38(2)			
$v_{LAO}(X_1)$	7.14(2)			
$v_{TO}(X_4)$	8.17(3)			
$v_{TA}(L_3)$	1.87(2)			
$v_{LA}(L_{2'})$	6.63(4)			
$v_{TO}(L_{3'})$	8.55(3)			
$v_{TO}(L_1)$	7.27(2)			
second order elastic moduli (in 10^{11} dyn cm^{-2}):				
c_{11}	12.40	$T = 298$ K	ultrasound measurements for temperature dependence, see Fig. 8	71B
c_{12}	4.13			
c_{44}	6.83			

third order elastic moduli: Fig. 9.

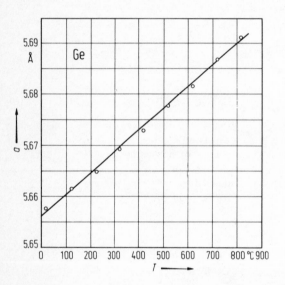

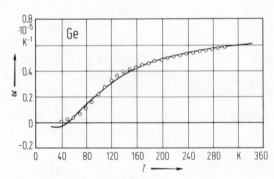

Fig. 5. Ge. Lattice parameter vs. temperature [68S].
◄

Fig. 6. Ge. Coefficient of linear thermal expansion vs. temperature in the range 40 to 360 K, single crystal [60N].

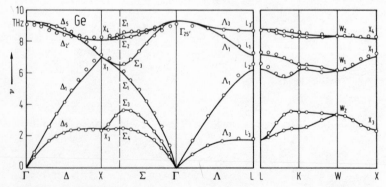

Fig. 7. Ge. Phonon dispersion relations. Experimental points from [71N, 72N1], solid lines: theory [77W].

Physical property	Numerical value	Experimental conditions	Experimental method, remarks	Ref.

Transport properties

Low field transport is maintained by electrons in the L_6 minima of the conduction band and holes near the point Γ_8 in the valence bands. At RT the mobility of samples with impurity concentrations below $10^{15}\,cm^{-3}$ is limited essentially by lattice scattering. At 77 K, even for doping concentrations below $10^{13}\,cm^{-3}$ the mobilities depend on the impurity concentration.

intrinsic conductivity:

σ_i	$2.1 \cdot 10^{-2}\,\Omega^{-1}\,cm^{-1}$	300 K	for temperature dependence, see Fig. 10	54M1

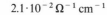

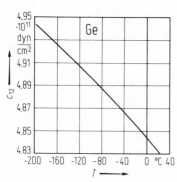
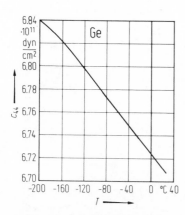

Fig. 8. Ge. Second order elastic moduli vs. temperature [53M].

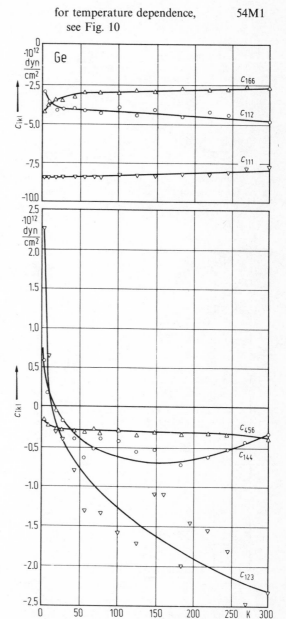

Fig. 9. Ge. Third-order elastic moduli vs. temperature. Solid lines: best fit to the data [83P].

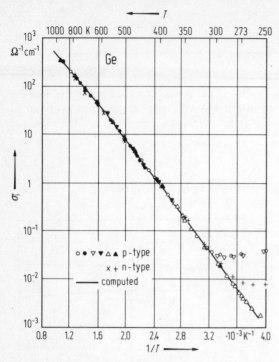

Fig. 10. Ge. Electrical conductivity vs. reciprocal temperature for six p-type and two n-type samples, solid line from empirical expressions for n_1 and the mobilities as given in the tables [54M1].

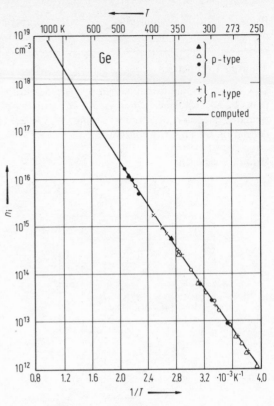

Fig. 11. Ge. Intrinsic carrier concentration vs. reciprocal temperature. Experimental data on four p-type and two n-type samples; solid line from empirical expression given in the tables [54M1].

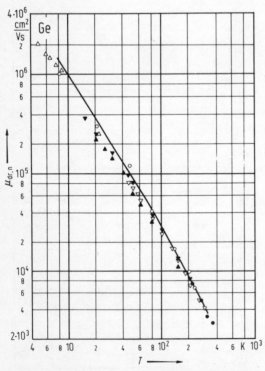

Fig. 12. Ge. (Ohmic) drift mobility of electrons vs. temperature obtained with a time-of-flight technique in hyperpure material (open circles); other symbols: data from five references; solid line: theory [81J].

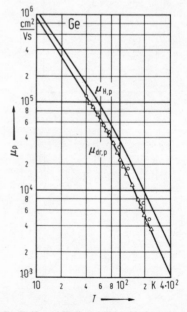

Fig. 13. Ge. Drift and Hall mobility of holes vs. temperature in the phonon-limited regime. Circles and triangles: experimental data, solid curves: theoretical analysis [83S].

Physical property	Numerical value	Experimental conditions	Experimental method, remarks	Ref.
intrinsic carrier concentration (in cm^{-3}):				
n_i	$2.33 \cdot 10^{13}$	300 K	for temperature dependence, see Fig. 11	54M1
	$1.76 \cdot 10^{16} T^{3/2}$ $\exp(-0.785/2\,kT)$		best fit to experimental data (kT in eV, T in K)	
electron mobility (in cm^2/Vs):				
μ_n	3900	300 K	lattice mobility ($n_d \approx 10^{14}\,cm^{-3}$) determined by drift experiments with injected minority carriers	53P
	$4.90 \cdot 10^7 \, T^{-1.66}$		best fit to experimental data in the range $77 \cdots 300$ K	54M1
For temperature dependence, see Fig. 12.				
$\mu_n(\Delta_6)$	$800\,cm^2/Vs$	300 K	Hall measurements under a pressure of 70 kbar at a sample with $n_d = 4.7 \cdot 10^{13}\,cm^{-3}$	79A
hole mobility (in cm^2/Vs):				
μ_p	1800	300 K	lattice mobility in high purity samples	54M
	$1.05 \cdot 10^9 \, T^{-2.33}$		best fit to experimental data in the range $100 \cdots 300$ K	
For temperature dependence, see Fig. 13.				
thermal conductivity:				
Temperature dependence: Fig. 14.				

Optical properties

refractive index, spectral dependence at RT:

n	4.00541 (11)	$\lambda = 8\,\mu m$	mean values for ten samples	82E1
	4.00412 (12)	$9\,\mu m$	from various suppliers	
	4.00319 (11)	$10\,\mu m$	measured at 20.0 °C	
	4.00248 (10)	$11\,\mu m$		
	4.00194 (11)	$12\,\mu m$	temperature coefficient	
	4.00151 (10)	$13\,\mu m$	between 20 °C and 25 °C: $4.0 \cdot 10^{-4}\,°C^{-1}$	

dielectric constant:

		$T[K], \nu[MHz]$		
ε	16.5	4.2, 750	capacitance bridge	66R
	16.0	4.2, 9200	microwave measurement	56A
	16.2	300 9200		

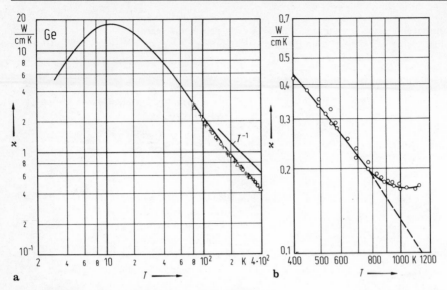

a

b

Fig. 14. Ge. Thermal conductivity vs. temperature. (a) 3···400 K, (b) 400···1200 K. Solid curve in (a) and data in (b) from [64G], experimental data in (a) from [84B]. Dashed line in (b): extrapolated lattice component.

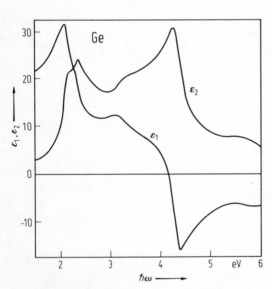

Fig. 15. Ge. Real and imaginary parts of the dielectric constant vs. photon energy [83A1].

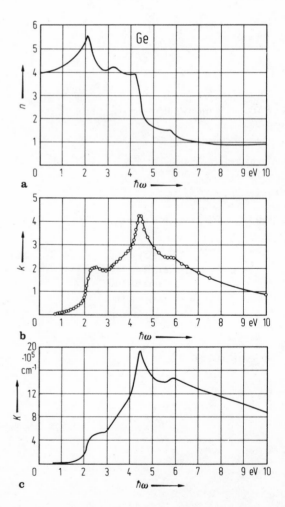

a

b

c

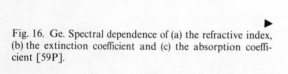

Fig. 16. Ge. Spectral dependence of (a) the refractive index, (b) the extinction coefficient and (c) the absorption coefficient [59P].

optical constants

real and imaginary parts of the dielectric constant measured by spectroscopical ellipso metry; n, k, R, K calculated from these data 83A1; see also Fig. 15.

$\hbar\omega$ [eV]	ε_1	ε_2	n	k	R	K [10^3 cm^{-1}]
1.5	21.560	2.772	4.653	0.298	0.419	45.30
2.0	30.361	10.427	5.588	0.933	0.495	189.12
2.5	13.153	20.695	4.340	2.384	0.492	604.15
3.0	12.065	17.514	4.082	2.145	0.463	652.25
3.5	9.052	21.442	4.020	2.667	0.502	946.01
4.0	4.123	26.056	3.905	3.336	0.556	1352.55
4.5	-14.655	16.782	1.953	4.297	0.713	1960.14
5.0	-8.277	8.911	1.394	3.197	0.650	1620.15
5.5	-6.176	7.842	1.380	2.842	0.598	1584.57
6.0	-6.648	5.672	1.023	2.774	0.653	1686.84

The spectral dependence of n, k, and K below 10 eV is shown in Fig. 16.

Impurities and defects

solubility of impurities in Ge

Element	c_{max}^{eq} cm^{-3}	T_{max} °C	T-range °C	Ref.
Al	$4.3 \cdot 10^{20}$	675	496 $\cdots$ 937	59T2
Ag	$4 \cdot 10^{18}$	850	770 $\cdots$ 930	62K2
As	$8.1 \cdot 10^{19}$	800	740 $\cdots$ 937	62T1
Au	$4.1 \cdot 10^{16}$	≈ 885	700 $\cdots$ 870	64K
Be	$4 \cdot 10^{16}$	937	720 $\cdots$ 920	61B
Bi	$6 \cdot 10^{16}$	910	867 $\cdots$ 937	62T1
Cd	$2 \cdot 10^{18}$	840	760 $\cdots$ 910	60K
Cu	$3.6 \cdot 10^{16}$	878	577 $\cdots$ 927	85S2
Fe	$1.5 \cdot 10^{15}$	850	750 $\cdots$ 940	57B
Ga	$4.9 \cdot 10^{20}$	670	303 $\cdots$ 937	59T2
H	$2 \cdot 10^{15}$	937		82H1
He	$6.5 \cdot 10^{12}$	937	795 $\cdots$ 872	56W
In	$4 \cdot 10^{18}$	800	300 $\cdots$ 937	60T
Li	$7.5 \cdot 10^{18}$	800	593 $\cdots$ 899	57P
Ni	$8.0 \cdot 10^{15}$	890	800 $\cdots$ 937	59T1
O	$1.3 \cdot 10^{18}$	≈ 937		56T1
P	$2 \cdot 10^{20}$	600	500 $\cdots$ 900	62A
Pb	$5.6 \cdot 10^{17}$	870		60T
Sb	$1.2 \cdot 10^{19}$	800	708 $\cdots$ 937	62T1
Sn	$5 \cdot 10^{20}$	≈ 300	400 $\cdots$ 937	60T
Te	$6.3 \cdot 10^{18}$	840	770 $\cdots$ 900	62I
Tl	$9.5 \cdot 10^{18}$	917	800 $\cdots$ 930	62T2
Zn	$2.5 \cdot 10^{18}$	770	420 $\cdots$ 760	60T

diffusion coefficients of impurities and self-atoms in Ge

Element	D_0 cm^2 s^{-1}	Q eV	T-range °C	D (800 °C) cm^2 s^{-1}	Ref.
Al	$1.0 \cdot 10^3$	3.45	554$\cdots$905	$6.2 \cdot 10^{-14}$	82D1
Ag	$4 \cdot 10^{-2}$	2.23	770$\cdots$920	$1.3 \cdot 10^{-12}$	62K2
As	10.3	2.49	700$\cdots$790	$2.1 \cdot 10^{-11}$	68I
Au	$2.5 \cdot 10^2$	2.3	600$\cdots$750	$3.9 \cdot 10^{-9}$	68G
B	$9.5 \cdot 10^6$	4.5	760$\cdots$850	$6.9 \cdot 10^{-15}$	67M
Be	0.5	2.5	720$\cdots$920	$9.1 \cdot 10^{-13}$	61B
Bi	3.3	2.5		$6.0 \cdot 10^{-12}$	59R
Cd	$1.75 \cdot 10^9$	4.42	760$\cdots$915	$3.0 \cdot 10^{-12}$	60K
Co	0.16	1.12	750$\cdots$850	$8.8 \cdot 10^{-7}$	61W
Cu	5.5	1.55	577$\cdots$927	$2.9 \cdot 10^{-7}$	85S2
Fe	0.13	1.1	775$\cdots$930	$8.9 \cdot 10^{-7}$	57B
Ga	$1.4 \cdot 10^2$	3.31	554$\cdots$916	$4.0 \cdot 10^{-14}$	86S
Ge	13.2	3.09	535$\cdots$904	$4.2 \cdot 10^{-14}$	85W1
H	$2.72 \cdot 10^{-3}$	0.38	800$\cdots$910	$4.5 \cdot 10^{-5}$	60F
In	$1.8 \cdot 10^4$	3.63	554$\cdots$919	$1.6 \cdot 10^{-13}$	82D2
Li	$1 \cdot 10^{-6}$	0.43	50$\cdots$150	$9.6 \cdot 10^{-9}$	76H
Na	0.029	1.55	660$\cdots$830	$1.5 \cdot 10^{-9}$	77S
Ni	0.8	0.91	700$\cdots$875	$4.3 \cdot 10^{-5}$	54M2
O	0.40	2.08	285$\cdots$760	$6.8 \cdot 10^{-11}$	64C
P	$3.3 \cdot 10^2$	3.1	600$\cdots$780	$9.1 \cdot 10^{-13}$	78M
Pb	$1.6 \cdot 10^3$	3.6	800	$2.0 \cdot 10^{-14}$	63B
Sb	21	2.08	650$\cdots$850	$3.6 \cdot 10^{-9}$	70D
Sn	70	3.05	798$\cdots$846	$5.7 \cdot 10^{-13}$	58V
Si	0.24	2.9	650$\cdots$900	$5.7 \cdot 10^{-15}$	81R
Te	2.0	2.82	770$\cdots$880	$2.1 \cdot 10^{-13}$	62I
Tl	$1.7 \cdot 10^3$	3.4	800$\cdots$930	$1.8 \cdot 10^{-13}$	62T2
Zn	0.65	2.55	825$\cdots$918	$6.9 \cdot 10^{-13}$	56K

binding energies of shallow impurities

Impurity	E_b[meV]	T[K]	Experimental method, remarks	Ref.
Group V substitutional donors				
Theory	9.81		donor effective mass calculation	69F
Sb	10.29	10, 4	optical absorption (photothermal ionization)	75S
Bi	12.81	10	optical absorption	64R
P	12.89	10	photoconductivity, absorption	74S
As	14.17	10	absorption	64R
Other shallow single donors				
Li	10.02	4	photoconductivity, absorption	72S
O-Li	10.47	6	piezophotoconductivity. Li tunnels between $4\langle 111 \rangle$ directions of molecular complex	78H
O-H	12.46	4	photoconductivity, labelled C in early work; obtained on quenching from 400 °C; isotope shift of 0.051 meV to lower energy on deuteration; tunnelling H center	80J
Group III substitutional acceptors				
Theory	11.2		acceptor EMT calculation	76B
B	10.82	4, 8	photoconductivity, optical absorption	74H

Impurity	E_b[meV]	T[K]	Experimental method, remarks	Ref.
Al	11.15		full series of EMT	
Ga	11.32		like excited states	
In	11.99		observed	
Tl	13.45		optical absorption	65J
Neutral double acceptors				
Be^0	24.80	2,8	optical absorption and/or photoconductivity	83C
Zn^0	32.98	2,8	optical absorption and/or photoconductivity	71M
Mg^0	35.85		optical absorption and/or photoconductivity	85M
Cd^0	54.96		optical absorption and/or photoconductivity	71M
Hg^0	91.65	6···20	optical absorption and/or photoconductivity	71M
Mn^0	55		optical absorption and/or photoconductivity	85M
Singly ionized double acceptors				
Be^-	58.02		optical absorption	83C
Zn^-	86.51	8···10	optical absorption, excited states observed	83C
Cd^-	160	20	Hall effect, photoconductivity	59T1
Hg^-	230	10	optical absorption	67C
Mn^-	100			85M
Group VI substitutional donors				
Se	286.2	19	photoconductivity	85G
	140, 280	10···300	Hall effect	59T1
Te	110	10···300	Hall effect	59N
	300			59T1
S	180, 280	10···300	Hall effect	59T1
	210, 250	60···160	DLTS	82P
O	17		Hall effect, photoconductivity	62K1
	40			82C
	200			
	60··· 80			78E
	160···180			
	16··· 17.3	7··· 40	absorption, D^0 state of oxygen-related thermal double donor	84C

energy levels of deep centers

Negative energy refers to conduction band edge, positive energy refers to valence band edge.

Impurity	E[meV]	T[K]	Remarks	Ref.
Group I impurities				
Cu^0	+43.25	20···300	Hall effect, photoconductivity	85S1
Cu^-	+330	20···300		84P
Cu^{2-}	−260, +410	20···300		84P
Au^0 acceptor	+160	20···300	Hall effect, photoconductivity	55D
Au^- acceptor	−200	20···300		57W
Au^{2-} acceptor	−40			65O
Au donor	+50		interstitial Au	
Ag^0	+130		Hall effect	59N
Ag^-	−280			
Ag^{2-}	−90			
Transition metal impurities				
Cr	+70, +120		resistivity	59N
Mn	+160, −370	30···400	resistivity, photoconductivity	59T1

40 References for 1.3

Impurity	E[meV]	T[K]	Remarks	Ref.
Fe	$+350, -270$	$30\cdots400$	resistivity, photoconductivity	59T1
Co	$+250, -300$	$77\cdots300$	resistivity, Hall effect and photoconductivity	59T1
Pt	$+40$		resistivity, three acceptor levels, interpretation not clear	59N
Pd	$+0.03, -0.18$		resistivity, Hall effect	80G

energy levels of defect centers

Defect	E[meV]	Generation	Remarks	Ref.
Di-vacancy (V_2)-donor			$E_v + 100, 120, 160$ meV levels are interpreted as di-vacancy-donor complexes and are double acceptors	77M
	$+100$	1 MeV	P, As doped Ge	
	$+120$	γ-irr. at 280 K	Sb doped Ge	
	$+160$		Bi doped Ge	
Acceptors labeled by annealing		γ-irr. at 42 K	(a) double acceptors, anneal at 65 K, interpreted as vacancy-interstitial (Frenkel) pair "65 K" defects.	77M
			(b) acceptors, anneal at $160\cdots200$ K, arise from interstitial defects observed in dislocation	
Di-vacancy (V_2)-H	$+80$ $+200$		free, high purity Ge grown in H-atmosphere; also produced by γ-irradiation of dislocated material. Attributed to di-vacancy-hydrogen complexes (c.f. di-vacancy-donor complexes above). $E_v + 71$ meV obtained if degeneracy factor of 4 assumed.	77H 82E2
Di-vacancy (V_2)-Li	$+100$	γ-irr. at 280 K	Li doped, high purity Ge; resistivity, Hall effect	77H

References for 1.3

52S   Straumanis, M.E., Aka, A.Z.: J. Appl. Phys. **23** (1952) 330.
53P   Prince, M.B.: Phys. Rev. **92** (1953) 681.
54M1   Morin, F.J., Maita, J.P.: Phys. Rev. **94** (1954) 1525.
54M2   van der Maesen, F., Brenkman, J.A.: Philips Res. Rep. **9** (1954) 225.
55D   Dunlap, W.C.: Phys. Rev. **97** (1955) 664.
56D   Dexter, R.N., Zeiger, H.J., Lax, B.: Phys. Rev. **104** (1956) 637.
56K   Kosenko, V.E.: Proc. Acad. Sci. USSR, Phys. Ser. (English Transl.) **20** (1956) 1399.
56T   Thurmond, C.D., Guldner W.G., Beach, A.L.: J. Electrochem. Soc. **103** (1956) 603.
56W   van Wieringen, A., Warmolitz, N.: Physica **22** (1956) 849.
57B   Bugai, A.A., Kosenko, V.E., Miselyuk, E.G.: Sov. Phys. Tech. Phys. (English Transl.) **2** (1957) 183.
57M   MacFarlane, G.G., McLean, T.P., Quarrington, J.E., Roberts, V.: Phys. Rev. **108** (1957) 1377.
57P   Pell, E.M.: J. Phys. Chem. Solids **3** (1957) 74.
57W   Woodbury, H.H., Tyler, W.W.: Phys. Rev. **105** (1957) 84.
58V   Valenta, M.W.: Ph. D. Thesis, Univ. Illinois **1958** (Univ. Microfilm 58-5509); Bull. Am. Phys. Soc. **2** (1958) 102.
59N   Newman, R., Tyler, W.W.: Solid State Physics, Vol. 8, Seitz, F., Turnbull, D. (eds.) New York: Academic Press **1959**, p. 49.
59P   Philipp, H.P., Taft, E.A.: Phys. Rev. **113** (1959) 1002.
59R   Reiss, H., Fuller, C.S.: Semiconductors, Hannay, N.B. (ed.), New York: Reinhold Publ. Corp., **1959**.
59T1   Tyler, W.W.: J. Phys. Chem. Solids **8** (1959) 59.
59T2   Trumbore, F.A., Probansky, E.M., Tartaglia, A.A.: J. Phys. Chem. Solids **11** (1959) 239.
59Z   Zwerdling, S., Lax, B., Roth, L.M., Button, K.J.: Phys. Rev. **114** (1959) 80.
60F   Frank, R.C., Thomas jr., J.E.: J. Phys. Chem. Solids **16** (1960) 144.
60K   Kosenko, V.E.: Sov. Phys. Solid State (English Transl.) **1** (1960) 1481.
60T   Trumbore, F.A.: Bell. Syst. Tech. J. **39** (1960) 205.
61B   Belyaev, Yu.I., Zhidkov, V.A.: Sov. Phys. Solid State (English Transl.) **3** (1961) 133.
61W   Wei, L.Y.: J. Phys. Chem. Solids **18** (1961) 162.
62A   Abrikosov, N. Kh., Glasov, V.M., Lin Chên-Yüan: Russ. J. Inorg. Chem. (English Transl.) **7** (1962) 429.

62K1   Kaiser, W.: J. Phys. Chem. Solids **23** (1962) 225.
62K2   Kosenko, V.E.: Sov. Phys. Solid State (English Transl.) **4** (1962) 42.
62I   Ignatkov, V.D., Kosenko, V.E.: Sov. Phys. Solid State (English Transl.) **4** (1962) 1193.
62T1   Trumbore, F.A., Spitzer, W.G., Logan, R.A., Luke, C.L.: J. Electrochem. Soc. **109** (1962) 734.
62T2   Tagirov, V.I., Kuliev, A.A.: Sov. Phys. Solid State (English Transl.) **4** (1962) 196.
63B   Boltaks, B.I.: Diffusion in Semiconductors, London: Infosearch Ltd., **1963**.
64C   Corbett, J.W., McDonald, R.S., Watkins, G.D.: J. Phys. Chem. Solids **25** (1964) 873.
64G   Glassbrenner, C.J., Slack G.A.: Phys. Rev. **134A** (1964) 1058.
64K   Kodera, H.: Jpn. J. Appl. Phys. **3** (1964) 369.
64R   Reuszer, J.H., Fisher, P.: Phys. Rev. **135** (1964) A1125.
65J   Jones, R.L., Fisher, P.: J. Phys. Chem. Solids **26** (1965) 1125.
65O   Ostroborodova, V.V.: Fiz. Tverd. Tela **7** (1965) 610; Sov. Phys.-Solid State (English Transl.) **7** (1965) 484.
67C   Chapman, R.A., Hutchinson, W.G.: Phys. Rev. **157** (1967) 615.
67M   Meer, W., Pommerrenig, D.: Z. Angew. Phys. **23** (1967) 369.
68G   Gromova, O.N., Khodunova, K.M.: Fiz. Khim. Obrab. Mater. **5** (1968) 150; Diffusion and Defect Data **3** (1969) 142.
68I   Isawa, N.: Jpn. J. Appl. Phys. **7** (1968) 81.
68S   Singh, H.P.: Acta Crystallogr. **24a** (1968) 469.
69A   Aggarwal, R.L., Zuteck, M.D., Lax, B.: Phys. Rev. **180** (1969) 800.
69F   Faulkner, R.A.: Phys. Rev. **184** (1969) 713.
70A   Aggarwal, R.L.: Phys. Rev. **B2** (1970) 446.
70D   Dudko, G.V., Marunina, N.I., Sukhov, G.V., Cherednichenko, D.I.: Sov. Phys. Solid State (English Transl.) **12** (1970) 1016.
71B   Burenkov, Yu.A., Nikanorov, S.P., Stepanov, A.V.: Sov. Phys. Solid State (English Transl.) **12** (1971) 1940; Fiz. Tverd. Tela **12** (1970) 2428.
71M   Moore, W.J.: J. Phys. Chem. Solids **32** (1971) 93.
71N   Nilsson, G., Nelin, G.: Phys. Rev. **B3** (1971) 364.
72N1   Nelin, G., Nilsson, G.: Phys. Rev. **B5** (1972) 3151.
72N2   Nilsson, G., Nelin, G.: Phys. Rev. **B6** (1972) 3777.
72S   Seccombe, S.D., Korn, D.: Solid State Commun. **11** (1972) 1539.
74H1   Haller, E.E., Hansen, W.L.: Solid State Commun. **15** (1974) 687.
74H2   Hensel, J.C., Suzuki, K.: Phys. Rev. **B9** (1974) 4219.
74S   Skolnick, M.S., Eaves, L., Stradling, R.A., Portal, J.C., Askenazy, S.: Solid State Commun. **125** (1974) 1403.
75B   Baker, J.F.C., Hart, M.: Acta Crystallogr. **31a** (1975) 2297.
75W   Wiley, J.D.: in "Semiconductors and Semimetals", Vol. 10, R.K. Willardson, A.C. Beer eds., Academic Press, New York **1975**.
75S   Skolnick, M.S., Eaves, L., unpublished.
76A   Altarelli, M., Lipari, N.O.: Phys. Rev. Lett. **36** (1976) 619.
76B   Baldereschi, A., Lipari, N.O.: Proc. 13th Int. Conf. on the Physics of Semicond., Rome 1976, Fumi, F.G. (ed.) Marves **1976**, p. 595.
76C   Chelikowsky, J.R., Cohen, M.L.: Phys. Rev. **B30** (1976) 556.
76F   Fink, D., Braunstein, R.: Phys. Status Solidi (b) **73** (1976) 361.
76H   Hufschmidt, M., Möller, W., Pfeiffer, T.: Vak.-Tech. **25** (1976) 206.
77H   Haller, E.E., Hubbard, G.S., Hansen, W.L.: IEEE Trans. Nucl. Sci. **NS24** (1977) 48.
77M   Mashovets, T.M.: Int. Conf. on Radiation Effects in Semiconductors, Dubrovnik 1976, Institute of Physics Conf. Ser. No. **31 1977**, p. 30.
77S   Stojić, M., Spirić, V., Kostoski, D.: Inst. Phys. Conf. Ser. **31** (1977) 304.
77W   Weber, W.: Phys. Rev. **B15** (1977) 4789.
78E   Emstev, V.V., Goncharev, L.A., Dostkhodzhoev, T.N.: Fiz. Tekh. Poluprovodn. **12** (1978) 139; Sov. Phys. Semicond. (English Transl.) **12** (1978) 78.
78H   Haller, E.E., Falicov, L.M.: Phys. Rev. Lett. **41** (1978) 1192.
78M   Matsumoto, S., Niimi, T.: J. Electrochem. Soc. **125** (1978) 1307.
79A   Ahmad, C.N., Adams, A.R., Pitt, G.D.: J. Phys. **C12** (1979) L379.
79M   Martin, T.P., Schaber, H.: Z. Physik **B35** (1979) 61.
80G   Golubev, N.F., Latyshev, A.V.: Sov. Phys. Semicond. (English Transl.) **14** (1980) 1074.
80J   Joos, B., Haller, E.E., Falicov, L.M.: Phys. Rev. **B22** (1980) 832.
80Z   Zverev, V.N.: Sov. Phys. Solid State (English Transl.) **22** (1980) 1921; Fiz. Tverd. Tela **22** (1980).
81J   Jacoboni, C., Nava, F., Canali, C., Ottaviani, G.: Phys. Rev. **B24** (1981) 1014.
81R   Räisänen, J., Hirvonen, J., Anttila, A.: Solid-State Electron. **24** (1981) 333.
82C   Clauws, P., Broeckx, J., Simoen, E., Vennik, J.: Solid State Commun. **44** (1982) 1011.
82D1   Dorner, P., Gust, W., Lodding, A., Odelius, H., Predel, B.: Acta Metall. **30** (1982) 941.
82D2   Dorner, P., Gust, W., Lodding, A., Odelius, H., Predel, B., Roll, U.: Z. Metallkd. **73** (1982) 325.
82E1   Edwin, R.P., Dudermel, M.T., Lamare, M.: Appl. Optics **21** (1982) 878.
82E2   Emstev, V.V., Mashovets, T.V., Nazaryan, E.K., Haller, E.E.: Sov. Phys. Semicond. (English Transl.) **16** (1982) 182.
82H1   Hansen, W.L., Haller, E.E., Luke, P.N.: IEEE Trans. Nucl. Sci. **NS-29** (1982) 738.
82H2   Hirose, Y., Shimomae, K., Hamaguchi, C.: J. Phys. Soc. Jpn. **51** (1982) 2226.
82P   Pearton, S.J.: Aust. J. Phys. **35** (1982) 53.
82W   Werner, A., Sanjorjo, J.A., Cardona, M.: Solid State Commun. **44** (1982) 155.
83A1   Aspnes, D.E., Studna, A.A.: Phys. Rev. **B27** (1983) 985.
83A2   Allen, P.B., Cardona, M.: Phys. Rev. **B27** (1983) 4760.

83C   Cross, J.W., Holt, L.T., Ramdas, A.K., Sauer, R., Haller, E.E.: Phys. Rev. **B28** (1983) 6953.
83L   López-Cruz, E., Cardona, M.: Solid State Commun. **45** (1983) 787.
83P   Philip, J., Breazeale, M.A.: J. Appl. Phys. **54** (1983) 752.
83Q   Qadri, S.B., Skelton, E.F., Webb, A.W.: J. Appl. Phys. **54** (1983) 3609.
83S   Szmulowicz, F.: Phys. Rev. **B28** (1983) 5943.
84B   Bakhchieva, S.R., Kekelidse, N.P., Kekua, M.G.: Phys. Status Solidi (a) **83** (1984) 139.
84C   Clauws, P., Vennik, J.: Phys. Rev. **B30** (1984) 4837.
84H   Hsieh, T.C., Miller, T., Chiang, T.C.: Phys. Rev. **B30** (1984) 7005.
84P   Pearton, S.J., Haller, E.E., Kahn, J.M.: J. Phys. **C17** (1984) 2375.
84V   Viña, L., Logothetidis, S., Cardona, M.: Phys. Rev. **B30** (1984) 1979.
85G   Grimmeiss, H.G., Larsson, K., Montelius, L.: Solid State Commun. **54** (1985) 863.
85L   Lautenschlager, P., Allen, P.B., Cardona, M.: Phys. Rev. **B31** (1985) 2163.
85M   McMurray, R.E.: Solid State Commun. **53** (1985) 1127.
85N   Nichols, J.M., Hansson, G.V., Karlsson, U.O., Persson, P.E.S., Uhrberg, R.I.G., Engelhard, R., Flodström, S.A., Koch,
      E.E.: Phys. Rev. **B32** (1985) 6663.
85S1 Salib, E.H., Fisher, P., Simmonds, P.E.: Phys. Rev. **B32** (1985) 2424.
85S2 Stolwijk, N.A., Frank, W., Hölzl, J., Pearton, S.J., Haller, E.E.: J. Appl. Phys. **57** (1985) 5211.
85W1 Werner, M., Mehrer, H., Hochheimer, H.D.: Phys. Rev. **B32** (1985) 3930.
85W2 Wachs, A.L., Miller, T., Hsieh, T.C., Shapiro, A.P., Chiang, T.C.: Phys. Rev. **B32** (1985) 2326.
86S   Sodervall, U., Odelius, H., Lodding, A., Roll. U., Predel, B., Gust. W., Dorner, P.: Philos. Mag. **A54** (1986) 539.

Physical property	Numerical value	Experimental conditions	Experimental method, remarks	Ref.

1.4 Grey tin (α-Sn)

Electronic properties

band structure: Fig. 1 (Brillouin zone, see Fig. 2 of section 1.1).

α-Sn is a *zero-gap semiconductor* with its lowest conduction band and its highest valence band being degenerate at Γ (symmetry Γ_8). A second conduction band with L_6-minima follows at a slightly higher energy. It determines the properties of n-type samples for $n > 10^{17}$ cm^{-3} ($T > 77$ K in intrinsic samples). Two further bands with Γ_7-symmetry, respectively, are situated below the Γ_8 valence band.

energies of symmetry points of the band structure (relative to the top of the valence band) (in eV):

$E(\Gamma_{6v})$	-11.34	non-local pseudopotential	76C
$E(\Gamma_{7v})$	-0.80	calculation (see Fig. 1)	
$E(\Gamma_{7c})$	-0.42		

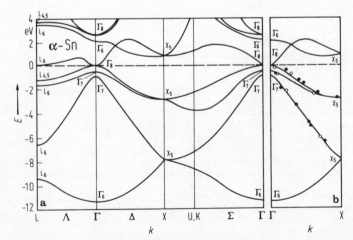

Fig. 1. α-Sn. (a) Band structure calculated by a non-local pseudopotential method [76C], (b) comparison with data from angular resolved photo-emission along the $\Gamma - X$ axis [83H].

Physical property	Numerical value	Experimental conditions	Experimental method, remarks	Ref.
$E(\Gamma_{8v,c})$	0.00		note that the Γ_{7c} conduction	
$E(\Gamma_{6c})$	2.08		band has shifted *below*	
$E(\Gamma_{8c})$	2.66		the Γ_{8v} band! Thus $E(\Gamma_{7c})$	
$E(X_{5v})$	-7.88		is negative.	
$E(X_{5v})$	-2.75			
$E(X_{5c})$	0.90			
$E(L_{6v})$	-9.44			
$E(L_{6v})$	-6.60			
$E(L_{6v})$	-1.68			
$E(L_{4,5v})$	-1.20			
$E(L_{6c})$	0.14			
$E(L_{6c})$	3.48			
$E(L_{4,5c})$	3.77			

energy differences above $\Gamma_{8v,c}$:

$E(L_{6c} - \Gamma_{8c})$	0.094 eV	0 K (extrapol.)	conductivity and Hall coefficient in the range 70⋯270 K, Fig. 2	56K
	0.12 eV		"optical band gap"	81F

energy differences below $\Gamma_{8v,c}$:

$E(\Gamma_{8v,c} - \Gamma_{7c})$	0.413 eV	1.5⋯85 K	interband magnetoreflection	70G
$\Delta_0(\Gamma_{8v,c} - \Gamma_{7v})$	0.8 eV			

critical point energies (in eV):
(measured by ellipsometry on InSb substrate stabilized layers)

E_1	1.316(5)	200 K	$\Lambda_{4,5v} - \Lambda_{6c}$	85V
$E_1 + \Delta_1$	1.798(6)		$\Lambda_{6v} - \Lambda_{6c}$	
E_0'	2.42(3)		at or near $\Gamma_{8v} - \Gamma_{6c}$	
$E_0' + \Delta_0'$	2.72(3)		$\Gamma_{8v} - \Gamma_{8c}$	
E_1'	4.28(4)		$L_{4,5v} - L_{6c}$	
$E_1' + \Delta_1'$	4.51(4)		$L_{4,5v} - L_{4,5c}$	

effective masses, electrons and holes (in units of m_0):

$m_{n,1}$	0.0236(2)	1.3 K	density of states mass	68B
$m_{n,h}$	0.21	4.2 K	$n(n_d)$-dependence	71L
$m_p(\Gamma_{8v})$	0.195		interband magnetoreflection	70G
$m_p(\Gamma_{7v})$	0.058			

valence band parameters:

A	15.0	1.3 K	Γ_{8c}-band	75L
B	22.9			
C^2	-696			

Lattice properties

Grey tin crystallizes in the diamond structure, space group O_h^7–Fd3m (α-modification). Slightly below RT it transforms into the metallic high-temperature β-modification (white tin). This transformation can be inhibited by alloying Ge or Si or by stabilizing grey tin as a heteroepitaxial layer on a substrate with a nearly equal lattice constant.

Physical property	Numerical value	Experimental conditions	Experimental method, remarks	Ref.
lattice parameter:				
a	6.4892(1) Å	20 °C	X-ray	59T
$\mathrm{d}a/\mathrm{d}T$	$3.1 \cdot 10^{-5}$ Å K^{-1}	$-130 \cdots 25$ °C		
transition temperature:				
T_{tr}	13.2(1) °C		equilibrium value	84V
	32.0(2) °C		hysteresis for the transformation in pure Sn	
linear thermal expansion coefficient: Fig. 3.				
density:				
d	7.285 g cm^{-3}	18 °C		60B
phonon dispersion relations: Fig. 4.				

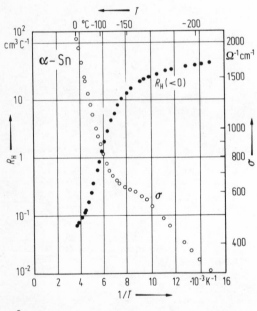

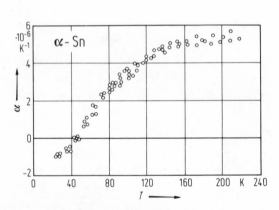

Fig. 2. α-Sn. Hall coefficient and conductivity vs. temperature of a sample containing about 10^{17} impurities/cm^3 [56K].

Fig. 3. α-Sn. Coefficient of linear thermal expansion vs. temperature [61N].

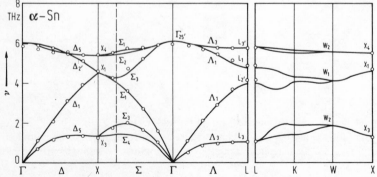

Fig. 4. α-Sn. Phonon dispersion curves. Experimental values from [71P], solid lines calculated with an adiabatic charge bond model [77W].

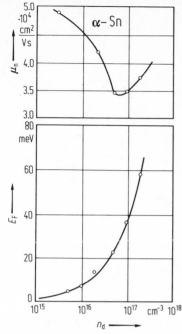

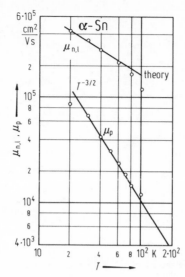

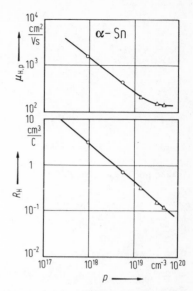

Fig. 5. α-Sn. Electron mobility at 5 K and Fermi energy vs. donor concentration. E_F measured from zero gap energy. An acceptor resonance level at 30 meV is suggested by the mobility minimum [82M].

Fig. 6. α-Sn. Light electron mobility and hole mobility vs. temperature [71L].

Fig. 7. α-Sn. Hall mobility (above) and Hall coefficient (below) vs. hole concentration at 77K. Data on bulk samples (open circles, [61T]) and substrate-stabilized metastable films [81F].

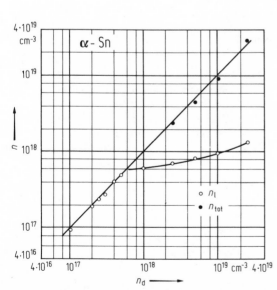

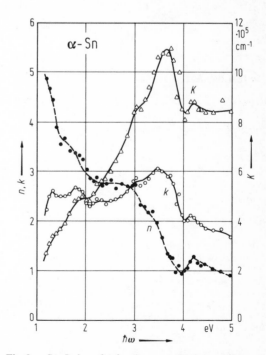

Fig. 8. α-Sn. Electron concentration at 4.2 K vs. donor concentration; n_1 (light electron concentration) from oscillatory magnetoresistance, n_{tot} (total electron concentration) from Hall coefficient [71L].

Fig. 9. α-Sn. Index of refraction n, extinction coefficient k and absorption coefficient K vs. photon energy [71H].

Physical property	Numerical value	Experimental conditions	Experimental method, remarks	Ref.
phonon frequencies (in THz):				
$\nu_{TO/LO}(\Gamma_{25'})$	6.00(6)	90 K	inelastic neutron scattering	71P
$\nu_{TA}(L_3)$	1.00(4)			
$\nu_{LA}(L_{2'})$	4.15(4)			
$\nu_{LO}(L_1)$	4.89(8)			
$\nu_{TO}(L_{3'})$	5.74(12)			
$\nu_{TA}(X_3)$	1.25(6)			
$\nu_{LA/LO}(X_1)$	4.67(6)			
$\nu_{TO}(X_4)$	5.51(8)			
second order elastic moduli (in 10^{12} dyn cm^{-2}):				
c_{11}	0.690		calculated from an eleven-	71P
c_{12}	0.293		parameter shell model fitted	
c_{44}	0.362		to experimental data	

Transport properties

In intrinsic samples light electrons and holes in the Γ_8-bands determine the transport properties. In heavily doped n-type samples the electrons in the L_6-minima dominate and grey tin behaves as an indirect gap semiconductor with E_g of about 0.09 eV (Fig. 2).

electron concentration: Fig. 8.

electron and hole mobilities (in cm^2/Vs):

$\mu_n(\Gamma_8)$	$1.2 \cdot 10^5$	100 K	from Hall data	71L
$\mu_n(L_6)$	$\approx 10^3$	4.2 K		
μ_p	$1 \cdot 10^4$	100 K	magnetoresistance for μ_n, μ_p, see also Figs. 5 ⋯ 7.	71L

Optical properties

Refractive index, extinction coefficient and absorption coefficient: Fig. 9.

dielectric constant:

ε	24	$T = 300$ K	background dielectric constant in the free carrier absorption region (from infrared reflectance measurements)	64L

References for 1.4

56K   Kohnke, E.E., Ewald, A.W.: Phys. Rev. **102** (1956) 1481.
59T   Thewlis, J., Davey, A.R.: Nature (London) **174** (1959) 1011.
60B   Busch, G.A., Kern, R.: Solid State Physics, F. Seitz, D. Turnbull eds., Academic Press, New York, Vol. 11, **1960**, p. 1.
61N   Novikova, S.I.: Sov. Phys. Solid State (English Transl.) **2** (1961) 2087; Fiz. Tverd. Tela **2** (1960) 2341.
61T   Tufte, O.N., Ewald, A.W.: Phys. Rev. **122** (1961) 1431.
64L   Lindquist, R.E., Ewald, A.W.: Phys. Rev. **135** (1964) A191.
68B   Booth, B.L., Ewald, A.W.: Phys. Rev. **168** (1968) 805.
70G   Groves, S.H., Pidgeon, C.R., Ewald, A.W., Wagner, R.J.: J. Phys. Chem. Solids **31** (1970) 2031.
71H   Hanyu, T.: J. Phys. Soc. Jpn. **31** (1971) 1738.
71L   Lavine, C.F., Ewald, A.W.: J. Phys. Chem. Solids **32** (1971) 1121.
71P   Price, D.L., Rowe, J.M., Nicklow, R.M.: Phys. Rev. **B3** (1971) 1268.
75L   Liu, L., Leung, W.: Phys. Rev. **B12** (1975) 2336.
76C   Chelikowsky, J.R., Cohen, M.L.: Phys. Rev. **B14** (1976) 556.
77W   Weber, W.: Phys. Rev. **B15** (1977) 4789.

81F   Farrow, R.F.C., Robertson, D.S., Williams, G.M., Cullis, A.G., Jones, G.R., Young, I.M., Dennis, P.N.J.,: J. Crystal
      Growth **54** (1981) 507.
82M   Myhra, S.: Phys. Status Solidi (b) **110** (1982) 97.
83H   Höchst, H., Hernández-Calderón, I.: Surf. Sci. **126** (1983) 25.
84V   Vnuk, F., DeMonte, A., Smith, R.W.: J. Appl. Phys. **55** (1984) 4171.
85V   Viña, L., Höchst, H., Cardona, M.: Phys. Rev. **B31** (1985) 958.

Physical property	Numerical value	Experimental conditions	Experimental method, remarks	Ref.

1.5 Silicon carbide (SiC)

Silicon carbide crystallizes in numerous (more than two hundred) different modifications (polytypes). The most important are:

cubic unit cell:

3C-SiC	space group: T_d^2–F$\overline{4}$3m	lattice: zincblende

hexagonal unit cell:

2H-SiC	space group: C_{6v}^4–P6$_3$mc	lattice: wurtzite

rhombohedral unit cell:

15R-SiC	space group: C_{3v}^5–R3m

Other polytypes with rhombohedral unit cell: 21R-SiC, 24R-SiC, 27R-SiC etc.

In all polytypes except 3C- and 2H-SiC atomic layers with cubic (c) and hexagonal (h) symmetry follow in a regular alternation in the direction of the c axis. This can be thought of as a natural one-dimensional superlattice imposed on the "pure" – i.e. h-layer free – 3C-SiC [77D1], the period of the superlattice being different for different modifications.

Electronic properties

3C-SiC:

band structure: Fig. 1 (Brillouin zone, see section 1.1, Fig. 2).

energy gaps (in eV):

$E_{g,ind}$ ($\Gamma_{15v} - X_{1c}$)	2.416(1)	2 K	wavelength modulated absorption	81B
$E_{g,dir}$ ($\Gamma_{15v} - \Gamma_{1c}$)	6.0	300 K	optical absorption	65D
E_{gx}	2.38807(3)	1.4 K	excitonic energy gap, wavelength modulated absorption	84G

According to [79R] the temperature dependence of E_{gx} can be described by the empirical formula $E_{gx} = 3.024 - 0.3055 \cdot 10^{-4} \, T^2/(311\,\text{K} - T)\,\text{eV}$. See also Fig. 5.

effective masses, electrons (in units of m_0):

$m_{n\parallel}$	0.677(15)	45 K	cyclotron resonance	85K
$m_{n\perp}$	0.247(11)			

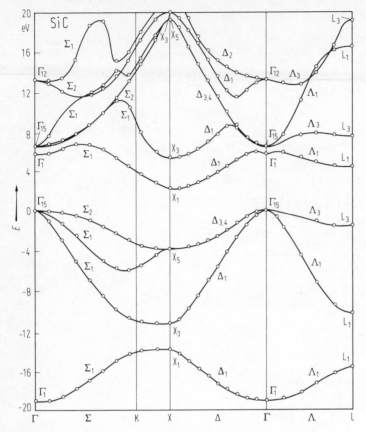

Fig. 1. SiC (3 C). Band structure [74H].

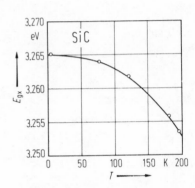

Fig. 3. SiC (4 H). Excitonic energy gap vs. temperature [64C].

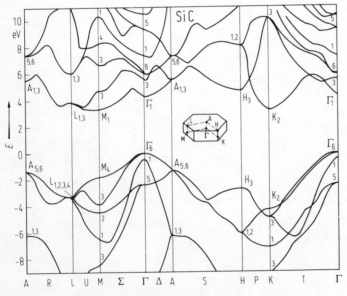

Fig. 2. SiC (2 H). Band structure [74H].

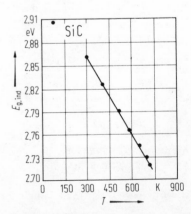

Fig. 4. SiC (6 H). Energy gap $E_{g,ind}$ vs. temperature [60P].

Physical property	Numerical value	Experimental conditions	Experimental method, remarks	Ref.

effective masses, holes:

Due to the small spin-orbit splitting the valence bands are highly non-parabolic, i.e. effective hole masses become strongly k-dependent.

2H-SiC:

band structure: Fig. 2.

E_{gx}	3.330 eV		optical absorption, Fig. 5	66P

4H-SiC:

E_{gx}	3.20 eV		see Figs. 3, 5	77D2

6H-SiC:

$E_{g,ind}$	2.86 eV	300 K	optical absorption, Figs. 4, 5	60P
E_{gx}	3.0230 eV		wavelength modulated absorption	81H
$m_{n\parallel}$	$1.5(2)m_0$	300 K	Faraday rotation	67E
$m_{n\perp}$	$0.25m_0$			
$m_{p,ds}$	$1.0m_0$			

8H-SiC:

E_{gx}	2.86 eV		see Fig. 5	77D2

15R-SiC:

E_{gx}	2.9863 eV		see Figs. 5, 6	81H
$m_{n\parallel}$	$0.53m_0$		Faraday rotation	67E
$m_{n\perp}$	$0.28(2)m_0$			

21R-SiC:

E_{gx}	2.92 eV		see Fig. 5	77D2

24R-SiC:

E_{gx}	2.80 eV		see Fig. 7	77D2

Lattice properties

3C-SiC:

lattice parameter:

a	4.3596 Å	297 K	Debye-Scherrer; for temperature dependence, see Fig. 8.	60T

linear thermal expansion coefficient:

α	$2.77(42)\cdot 10^{-6}\,K^{-1}$	300 K	recommended value	75S

melting point:

T_m	3103(40) K	$p = 35$ bar	peritectic decomposition temperature	60S

density:

d	$3.166\,g\,cm^{-3}$	293 K		69K

Physical property	Numerical value	Experimental conditions	Experimental method, remarks	Ref.

phonon dispersion relations: Fig. 9.

phonon wavenumbers (in cm^{-1}):

$\bar{\nu}_{TO}(\Gamma)$	796.2(3)	RT	Raman spectroscopy	82O1
$\bar{\nu}_{LO}(\Gamma)$	972.2(3)			
$\bar{\nu}_{TA}(L)$	266	RT	discussion of Raman spectra including data from previous publications	82O2
$\bar{\nu}_{LA}(L)$	610			
$\bar{\nu}_{TO}(L)$	766			
$\bar{\nu}_{LO}(L)$	838			
$\bar{\nu}_{TA}(X)$	373			
$\bar{\nu}_{LA}(X)$	640			
$\bar{\nu}_{TO}(X)$	761			
$\bar{\nu}_{LO}(X)$	829			

second order elastic moduli (in $10^{12}\,dyn\,cm^{-2}$):

c_{11}	3.523			75K1
c_{12}	1.404			
c_{44}	2.329			

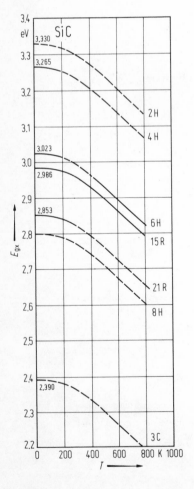

◀ Fig. 5. SiC. Excitonic energy gap of several polytypes vs. temperature [68C].

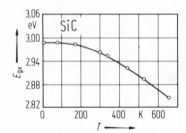

Fig. 6. SiC (15 R). Excitonic energy gap vs. temperature [63P].

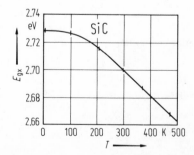

Fig. 7. SiC (24 R). Excitonic energy gap vs. temperature [64Z].

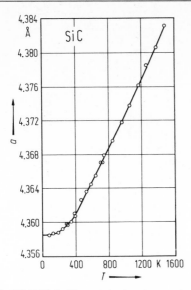

Fig. 8. SiC (3 C). Lattice constant vs. temperature [60T].

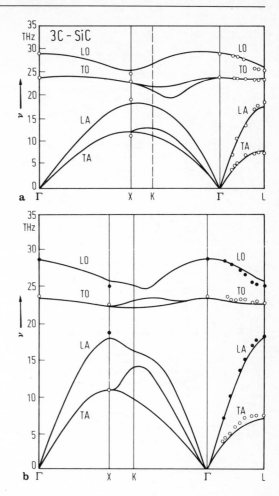

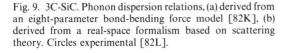

Fig. 9. 3C-SiC. Phonon dispersion relations, (a) derived from an eight-parameter bond-bending force model [82K], (b) derived from a real-space formalism based on scattering theory. Circles experimental [82L].

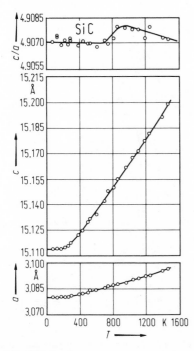

Fig. 10. SiC (6 H). Lattice constants vs. temperature [60T].

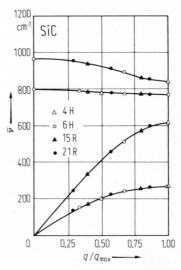

Fig. 11. SiC. Phonon dispersion relations for 4H, 6H, 15R and 21R polytypes along the axial direction vs. reduced wavevector [68F].

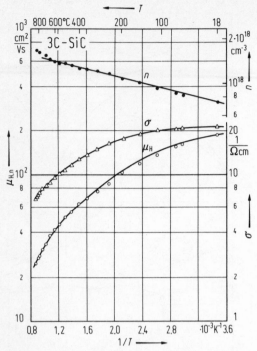

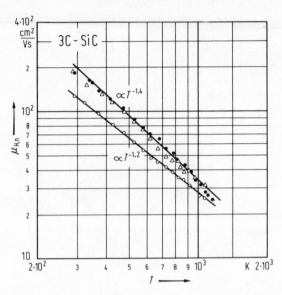

Fig. 12. 3C-SiC. Conductivity, carrier concentration and electron Hall mobility of an epitaxial layer on Si vs. temperature [84S].

Fig. 13. 3C-SiC. Electron Hall mobility of three expitaxial layers on Si vs. temperature [84S].

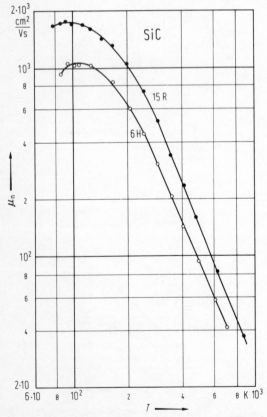

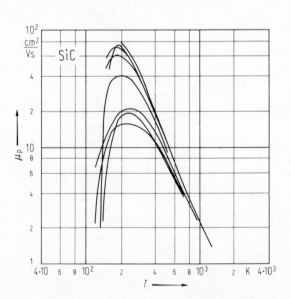

Fig. 14. SiC. Electron mobility vs. temperature in 6 H and 15 R samples [67B].

Fig. 15. SiC (6 H). Hole mobility of different samples vs. temperature [63V].

Physical property	Numerical value	Experimental conditions	Experimental method, remarks	Ref.

Hexagonal polytypes

6H-SiC:

lattice parameters:

a	3.0806 Å	297 K	Debye-Scherrer; for	60T
c	15.1173 Å		temperature dependence, see Fig. 10	

density:

d	$3.211 \, \text{g cm}^{-3}$	300 K		67G

phonon dispersion relations (also for 4H-, 15R-, and 21R-SiC): Fig. 11.

phonon energies (in meV):

$h\nu_{TA}$	36.3		discussion of free exciton	81H
	46.3		replica in wavelength	
$h\nu_{LA}$	53.3		modulated absorption	
	77.0			
$h\nu_{TO}$	95.6			
$h\nu_{LO}$	104.2			
	107.0			

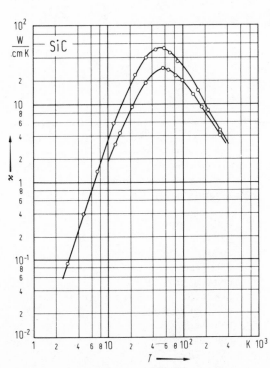

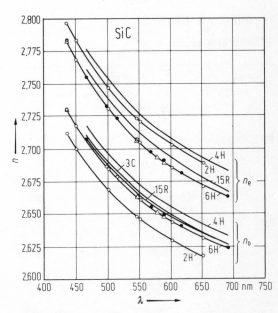

Fig. 16. SiC (6H). Thermal conductivity ($\perp c$-axis) vs. temperature for two different samples [64S].

Fig. 17. SiC. Dispersion of refractive indices of several polytypes [72P].

Physical property	Numerical value	Experimental conditions	Experimental method, remarks	Ref.

second order elastic moduli (in 10^{12} dyn cm^{-2}):

c_{11}	5.00 (20)	300 K	ultrasonic resonance	65A
c_{12}	0.92 (28)			
c_{33}	5.64 (2)			
c_{44}	1.68 (2)			

Transport and optical properties

electron and hole mobilities of 3C-SiC (in cm^2/Vs):

μ_n	900	300 K	crystalline	66N
	380	RT		83N
μ_p	15 $\cdots$ 21			83N

For Hall mobility in epitaxial layers on Si, see Figs. 12 and 13.
  Mobilities in other polytypes are of the same order of magnitude, see Figs. 14 and 15.

thermal conductivity of 6H-SiC: Fig. 16.

refractive index:
Empirical formula for 3C-SiC: $n(\lambda) = 2.55378 + 3.417 \cdot 10^4 \lambda^{-2} (\lambda = 467 \cdots 691$ nm) [69S]. For other polytypes see Fig. 17.

dielectric constants:

$\varepsilon(0)$	9.72	300 K	3C-SiC	70P
$\varepsilon(\infty)$	6.52			
$\varepsilon_\perp(0)$	9.66		6H-SiC	
$\varepsilon_\parallel(0)$	10.03			
$\varepsilon_\perp(\infty)$	6.52			
$\varepsilon_\parallel(\infty)$	6.70			

Impurities and defects

solubility of impurities in SiC

Impurity	c_{max}^{eq} [cm^{-3}]	T [°C]	Remarks	Ref.
N	$2.6 \cdot 10^{20}$	2450	crystal growth; nitrogen pressure 35 atm	65S
Be	$7 \cdot 10^{17}$	1800	diffusion experiments	68M
	$5 \cdot 10^{19}$	2300		
B	$2 \cdot 10^{19}$	1800	epitaxial growth; (0001) face	77T
	$1.5 \cdot 10^{20}$	2300		
	$4 \cdot 10^{19}$	1800	epitaxial growth; (000$\bar{1}$) face	77T
	$2.5 \cdot 10^{20}$	2300		
Al	$1.1 \cdot 10^{21}$	2300	epitaxial growth; (0001) face	77T
	$7 \cdot 10^{20}$	2300	epitaxial growth; (000$\bar{1}$) face	77T
Ga	$1.2 \cdot 10^{19}$	1800	epitaxial growth; (0001) face	77T
	$1.2 \cdot 10^{19}$	2300		
	$2.8 \cdot 10^{18}$	1800	epitaxial growth (000$\bar{1}$) face	77T
	$7 \cdot 10^{18}$	2300		

diffusion constants of impurities in 6H SiC

Dopant	D_0 cm^2 s^{-1}	Q eV	T °C	Remarks	Ref.
N	$4.6 \cdots 8.7 \cdot 10^{-4}$	$7.6 \cdots 9.3$	$2000 \cdots 2550$	pn-junction	66K
Be	32	5.2	$1950 \cdots 2250$	pn-junction, probably substitutional diffusion	68M
	0.3	3.1	$1700 \cdots 2100$	pn-junction, probably interstitial diffusion	68M
B	$1.6 \cdot 10^2$	5.6	$1800 \cdots 2250$	pn-junction	66V
	3.2	5.1	$1600 \cdots 2550$	pn-junction	74V
Al	1.8	4.9	$1700 \cdots 2000$	pn-junction	60C
	0.2	4.9	$1800 \cdots 2250$	pn-junction	66V
	8.0	6.1	$1900 \cdots 2300$	pn-junction	69M
	$1.3 \cdot 10^{-8}$	2.4	$1350 \cdots 1800$	Ion implantation; SIMS measurements	82T
Ga	0.17	5.5	$2050 \cdots 2300$	pn-junction	74V

binding energy of impurities in SiC

Due to the existence of inequivalent lattice sites in silicon carbide (except for the 3C and 2H polytypes), there are several site-dependent energy levels for each donor or acceptor, respectively. Under most experimental conditions, however, one can only distinguish between energy levels assigned to impurity atoms at c-sites and energy levels caused by impurity atoms at h-sites. Impurity atoms at c-sites generally exhibit deeper levels than those at h-sites.

A. Donor levels

Impurity	E_b [meV]	T [K]	Remarks	Ref.
N	53.6(5)	6	3C, photoluminescence	77D1
	56.5	4	3C, photoluminescence	76K
	66	4	4H, h-site, luminescence of donor-acceptor pairs	80I
	124	4	4H, c-site, luminescence of donor-acceptor pairs	80I
	55(7)	77	4H, photoluminescence	77S
	95	$60 \cdots 1000$	6H, $n_d = 5 \cdot 10^{16}$ cm^{-3}; evaluation of Hall data; $n(T)$	70H
	100	4	6H, h-site, luminescence of donor-acceptor pairs	80I
	155	4	6H, c-site, luminescence of donor-acceptor pairs	80I
	52	$60 \cdots 1000$	15R, $n_d = 3 \cdot 10^{16}$ cm^{-3}; evaluation of Hall data; $n(T)$	70H
	64	4	15R, h-site, luminescence of donor-acceptor pairs	80I
	112	4	15R, c-site, luminescence of donor acceptor pairs	80I

B. Acceptor levels

Impurity	E_b [meV]	T [K]	Remarks	Ref.
Be	320	$280 \cdots 420$	6H, photoluminescence	67K
	420	$300 \cdots 1000$	6H, temperature dependence of Hall data	68M
	600			
B	735	4	3C, photoluminescence	75K2
	647	77	4H, photoluminescence	80I
	698	77	6H, h-site, photoluminescence	80I
	723	77	6H, c-site, photoluminescence	80I
	666	77	15R, h-site, photoluminescence	80I
	700	77	15R, c-site, photoluminescence	80I

Impurity	E_b[meV]	T[K]	Remarks	Ref.
Al	216	1.8	3C, luminescence of donor-acceptor pairs	70C
	260	77	3C, photoluminescence	68Z
	254	4	3C, luminescence of donor-acceptor pairs	80I
	191	4	4H, luminescence of donor-acceptor pairs	80I
	239	4	6H, h-site, luminescence of donor-acceptor pairs	80I
	249	4	6H, c-site, luminescence of donor-acceptor pairs	80I
	280	160	6H, DLTS	85A
	315	300···1300	6H, $n_a = 5 \cdot 10^{18}$ cm^{-3}; evaluation of Hall data; $p(T)$	63V
	220	300···1300	6H, $n_a = 2 \cdot 10^{19}$ cm^{-3}; evaluation of Hall data; $p(T)$	63V
	206	4	15R, h-sites, luminescence of donor-acceptor pairs	80I
	221			
	223	4	15R, c-sites, luminescence of donor-acceptor pairs	80I
	230			
	236			
Ga	343	4	3C, photoluminescence	76K
	267	4	4H, luminescence of donor-acceptor pairs	80I
	317	4	6H, h-site, luminescence of donor-acceptor pairs	80I
	333	4	6H, c-site, luminescence of donor-acceptor pairs	80I
	282	4	15R, h-sites, luminescence of donor-acceptor pairs	80I
	300			
	305	4	15R, c-sites, luminescence of donor-acceptor pairs	80I
	311			
	320			

References for 1.5

60C   Chang, H.C., Le May, Ch.Z., Wallace, L.F.: Silicon carbide – a high temperature semiconductor, O'Connor, J.R., Smiltens, J. (eds.), Oxford, London, New York, Paris: Pergamon Press **1960**, p. 496.
60P   Philipp, H.R., Taft, E.A.: Silicon Carbide – A High Temperature Semiconductor, J.R. O'Connor and J. Smiltens eds., Pergamon Press, Oxford, London, New York, Paris **1960**, p. 366.
60S   Scace, R.I., Slack, G.A.: Silicon Carbide – A High Temperature Semiconductor, J.R. O'Connor and J. Smiltens eds., Pergamon Press, Oxford, London, New York, Paris **1960**, p. 24.
60T   Taylor, A., Jones, R.M.: Silicon Carbide – A High Temperature Semiconductor, J.R. O'Connor and J. Smiltjens eds., Pergamon Press, Oxford, London, New York, Paris **1960**, p. 147.
63P   Patrick, L., Hamilton, D.R., Choyke, W.J.: Phys. Rev. **132** (1963) 2023.
63V   Van Daal, H.J., Knippenberg, W.F., Wasscher, J.D.: J. Phys. Chem. Solids **24** (1983) 109.
64C   Choyke, W.J., Patrick, L., Hamilton, D.R.: Proc. 7th Int. Conf. Semicond., Paris 1964, M. Hulin ed., Dunod, Paris **1964**, p. 751.
64S   Slack, G.A.: J. Appl. Phys. **35** (1964) 3460.
64Z   Zanmarchi, G.: Proc. 7th Int. Conf. Semicond., Paris 1964, M. Hulin ed., Dunod, Paris **1964**, p. 57.
65A   Arlt, G., Schodder, G.R.: J. Acoust. Soc. Am. **37** (1965) 384.
65D   Dalven, R.: J. Phys. Chem. Solids **26** (1965) 439.
65S   Scace, R.I., Slack, G.A.: J. Chem. Phys. **42** (1965) 805.
66K   Kroko, L.J., Milnes, A.G.: Solid-State Electron. **9** (1966) 1125.
66N   Nelson, W.E., Holder, F.A., Rosenbloom, A.: J. Appl. Phys. **37** (1966) 333.
66P   Patrick, L., Hamilton, D.R., Choyke, W.J.: Phys. Rev. **143** (1966) 526.
66V   Vodakov, Yu.A., Mokhov, E.N., Reifman, M.B.: Sov. Phys.-Solid State **8** (1966) 1040.
67B   Barrett, D.L., Campbell, R.B.: J. Appl. Phys. **38** (1967) 53.
67E   Ellis, B., Moss, T.S.: Proc. Royal Soc. (London) **A299** (1967) 383, 393.
67G   Gomes de Mesquita, A.H.: Acta Crystallogr. **23** (1967) 610.
67K   Kalnin, A.A., Pasynkov, V.V., Tairov, Y.M., Yaskov, D.A.: Sov. Phys.-Solid State **8** (1967) 2381.
68C   Choyke, W.J.: Mater. Res. Bull. **4** (1969) S 141.
68F   Feldman, D.W., Parker, J.H., Choyke, W.J., Patrick, L.: Phys. Rev. **173** (1968) 787.
68M   Maslakovets, Yu.P., Mokhov, E.N., Vodakov, Yu.A., Lomakina, G.A.: Sov. Phys.-Solid State **10** (1968) 634.
68Z   Zanmarchi, G.: J. Phys. Chem. Solids **29** (1968) 1727.
69K   Kern, E.L., Hamill, D.W., Deem, H.W., Sheets, H.D.: Mater. Res. Bull. **4** (1969) 25.
69M   Mokhov, E.N., Vodakov, Yu.A., Lomakina, G.A.: Sov. Phys. Solid-State **11** (1969) 415.

69S   Shaffer, P.T.B., Naum, R.G.: J. Opt. Soc. Am. **59** (1969) 1498.

70C   Choyke, W.J., Patrick, L.: Phys. Rev. **B2** (1970) 4959.

70H   Hagen, S.H., Kapteyns, C.J.: Philips Res. Repts. **25** (1970) 1.

70P   Patrick, L., Choyke, W.J.: Phys. Rev. **B2** (1970) 2255.

72P   Powell, J.A.: J. Opt. Soc. Am. **62** (1972) 341.

74H   Hemstreet, L.A., Fong, C.Y.: Silicon Carbide – 1973, R.C. Marshall, J.W. Faust, C.E. Ryan eds., Univ. of South Carolina Press, Columbia, S.C. **1974**, p. 284.

74V   Vodakov, Yu.A., Mokhov, E.N.: Silicon Carbide, 1973, Marshall, R.C., Faust, J.W., Ryan, C.E. (eds.), Columbia, S.C.: University of South Carolina Press **1974**, p. 508.

75K1   Kunc, K., Balkanski, M., Nusimovici, M.A.: Phys. Status Solidi (b) **72** (1975) 229.

75K2   Kuwabara, H., Yamada, S.: Phys. Status Solidi **A30** (1975) 739.

75S   Slack, G.A., Bartram, S.F.: J. Appl. Phys. **46** (1975) 89.

76K   Kuwabara, H., Yamanaka, K., Yamada, S.: Phys. Status Solidi **A37** (1976) K 157.

77D1   Dean, P.J., Choyke, W.J., Patrick, L.: J. Lumin. **15** (1977) 299.

77D2   Dubrovskii, G.B., Lepneva, A.A.: Sov. Phys. Solid State (English Transl.) **19** (1977) 729; Fiz. Tverd. Tela **19** (1977) 1252.

77S   Suzuki, A., Matsunami, H., Tanaka, T.: J. Electrochem. Soc. **124** (1977) 241.

77T   Tairov, Y.M., Vodakov, Y.A.: Topics in Applied Physics, vol. 17: Electroluminescence, Pankove, J.I. (ed.), Berlin, Heidelberg, New York: Springer-Verlag **1977**, p. 31.

79R   Ravindra, N.M., Srivastava, V.K.: Phys. Chem. Solids **40** (1979) 791.

80I   Ikeda, M., Matsunami, H., Tanaka, T.: Phys. Rev. **B22** (1980) 2842.

81B   Bimberg, D., Altarelli, M., Lipari, N.O.: Solid State Commun. **40** (1981) 437.

81H   Humphreys, R.G., Bimberg, D., Choyke, W.J.: Solid State Commun. **39** (1981) 163.

82K   Kushawa, M.S.: Phys. Status Solidi (b) **111** (1982) 337.

82L   Lee, D.H., Joannopoulos, J.D.: Phys. Rev. Lett. **48** (1982) 1846.

82O1   Olego, D., Cardona, M., Vogl, P.: Phys. Rev. **B25** (1982) 3878.

82O2   Olego, D., Cardona, M.: Phys. Rev. **B25** (1982) 1151.

82T   Tajima, Y., Kijima, K., Kingery, W.D.: J. Chem. Phys. **77** (1982) 2592.

83N   Nishino, S., Powell, J.A., Will, H.A.: Appl. Phys. Lett. **42** (1983) 460.

84G   Gorban', I.S., Gubanov, V.A., Lysenko, V.G., Pletyushkin, A.A., Timofeev, V.B.: Sov. Phys. Solid State (English Transl.) **26** (1984) 1385; Fiz. Tverd. Tela **26** (1984) 2282.

84S   Sasaki, K., Sakuma, E., Misawa, S., Yoshida, S., Gonda, S.: Appl. Phys. Lett. **45** (1984) 72.

85A   Anikin, M.M., Lebedev, A.A., Syrkin, A.L., Suvorov, A.V.: Sov. Phys.-Semicond. **19** (1985) 69.

85K   Kaplan, R., Wagner, R.J., Kim, H.J., Davis, R.F.: Solid State Commun. **55** (1985) 67.

Physical property	Numerical value	Experimental conditions	Experimental method, remarks	Ref.

1.6 Silicon germanium alloys (Si_xGe_{1-x})

Silicon and germanium form a continuous series of solid solutions with gradually varying parameters.

The *band structure* shows a cross-over in the lowest conduction band edge from Ge-like [111] symmetry to Si-like [100] symmetry at $x \approx 0.15$ (Fig. 1). According to [83K] this value should lie a little bit higher ($x \cong 0.25$).

Compositional dependence of band gaps (calculated):

$E_g(\Gamma - X)$	$0.8941 + 0.0421\,x + 0.1691\,x^2$	85K
$E_g(\Gamma - L)$	$0.7596 + 1.0860\,x + 0.3306\,x^2$	

Transport properties have been mostly investigated on single crystals. The mobility is influenced by alloy scattering which contributes according to $\mu_{alloy} \propto T^{0.8}\,x^{-1}(1-x)^{-1}$. Near the band crossover ($x \approx 0.15$) intervalley scattering has to be taken into account.

Figures on transport properties: Figs. 2⋯5.

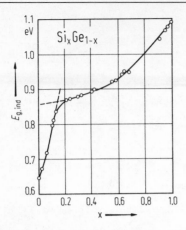

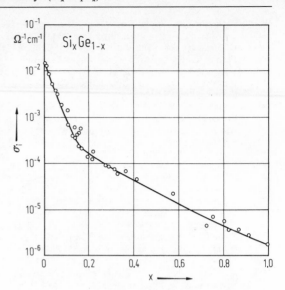

Fig. 1. Si$_x$Ge$_{1-x}$. Compositional dependence of the indirect energy gap at 296 K based on a one-phonon analysis of the absorption edge. At about x = 0.15 a crossover occurs of the Ge-like [111] conduction band minima and the Si-like [100] conduction band minima [58B].

Fig. 2. Si$_x$Ge$_{1-x}$. Composition dependence of the intrinsic conductivity at room temperature [60B].

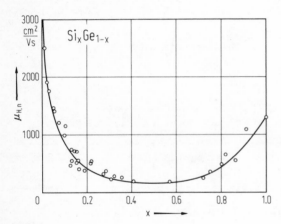

Fig. 3. Si$_x$Ge$_{1-x}$. Composition dependence of the electron Hall mobility at room temperature [60B].

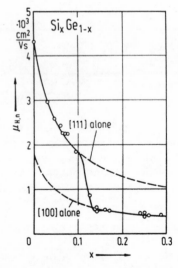

Fig. 4. Si$_x$Ge$_{1-x}$. Composition dependence of the electron Hall mobility for x < 0.3. The dashed curves show the calculated mobilities for electrons in [100]- and [111]-valleys. The solid curve takes both types of valleys into account and includes an arbitrary form of intervalley scattering [58G].

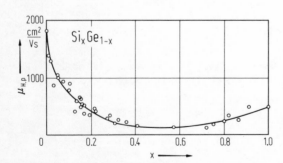

Fig. 5. Si$_x$Ge$_{1-x}$. Composition dependence of the hole Hall mobility at room temperature [60B].

References for 1.6

58B   Braunstein, R., Moore, A.R., Herman, F.: Phys. Rev. **109** (1958) 695.
58G   Glicksman, M.: Phys. Rev. **111** (1958) 125.
60B   Busch, G., Vogt, O.: Helv. Phys. Acta **33** (1960) 437.
83K   Kustov, E.F., Mel'nikov, E.A., Sutchenkov, A.A., Levadnii, A.I., Filikov, V.A.: Sov. Phys. Semicond. (English Transl.)
      17 (1983) 481; Fiz. Tekh. Poluprovodn. **17** (1983) 769.
85K   Krishnamurti, S., Sher, A., Chen, A.: Appl. Phys. Lett. **47** (1985) 160.

2 III–V compounds

Physical property	Numerical value	Experimental conditions	Experimental method, remarks	Ref.

2.1 Boron nitride (BN)

Boron nitride crystallizes in three modifications: BN_{cub} (cubic BN, zincblende structure), BN_{hex} (hexagonal BN) and BN_w (wurtzite structure). BN_{hex} is stable under normal conditions, BN_{cub} is metastable under normal conditions. The BN_w phase is metastable under all conditions.

2.1.A Cubic boron nitride

Electronic properties

All **band structure** calculations lead to an indirect gap structure, the conduction band minima being situated at X. Fig. 1 shows two recent calculations. (Brillouin zone: Fig. 2 of section 1.1.)

energy gaps (in eV):

$E_{g,ind}$	6.4 (5)	RT	ultraviolet absorption; other literature data lie in the range $6 \cdots 8\,eV$	74C
$(\Gamma_{15v} - X_{1c})$				
	6.99		calculated, Fig. 1b	85H
	8.6		calculated, Fig. 1a	84P
$E_{g,dir}$	14.5	RT	reflectivity	62P
$(\Gamma_{15v} - \Gamma_{1c})$	10.86		calculated, Fig. 1a	84P
	9.94		calculated, Fig. 1b	85H

effective masses (in units of m_0):

m_n	0.752		calculated from band structure data	85H
$m_{p,h}$	0.375	$\parallel[100]$		
	0.926	$\parallel[111]$		
$m_{p,l}$	0.150	$\parallel[100]$		
	0.108	$\parallel[111]$		

Lattice properties

lattice parameter:

a	3.6157 (10) Å		X-ray	74S

melting point:

T_m	2973 °C			57W

density:

d	3.4870 g cm^{-3}		X-ray	74S

phonon wavenumbers:

$\bar{v}_{LO}$	1305 (1) cm^{-1}	RT	Raman scattering	83S1
$\bar{v}_{TO}$	1054.7 (6) cm^{-1}			

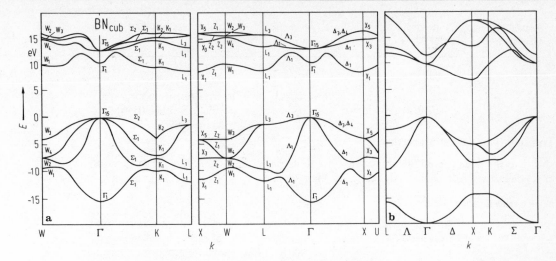

Fig. 1. BN, cubic. Band structure calculated with the LCAO-method, (a) including ionicity and fitting of APW results at high symmetry points [84P], (b) ab intitio calculation [84H].

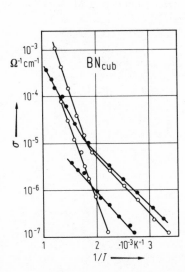

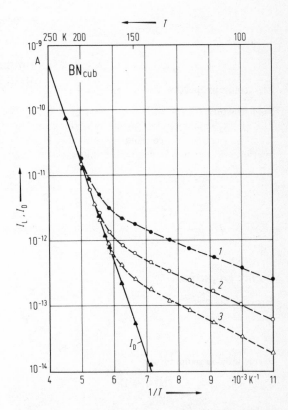

Fig. 2. BN_{cub}. Electrical conductivity vs. reciprocal temperature for different polycrystalline samples above RT [76B].

Fig. 3. BN_{cub}. Dark current (I_D) and photocurrent (I_L) at various illuminations (curve 1: $I = I_0$, 2: $I = 0.13 \, I_0$, 3: $I = 0.01 \, I_0$) vs. temperature for a small undoped single crystal [75H].

Physical property	Numerical value	Experimental conditions	Experimental method, remarks	Ref.

second order elastic moduli:

c_{11}	7.120 $\cdot 10^{12}\,\mathrm{dyn\,cm^{-2}}$	$T = 300\,\mathrm{K}$	interpolated from measured values of other III–V compounds	63S

Transport and optical properties

Owing to the large gap all literature data refer to extrinsic conduction. For the temperature dependence of dark conductivity and photoconductivity, see Figs. 2 and 3.

carrier concentration and mobility:

n	$10^{15}\,\mathrm{cm^{-3}}$	$T = 500\,\mathrm{K}$	polycrystalline material, type of carrier not determined	76B
μ	$0.2\,\mathrm{cm^2/Vs}$			
n	$10^{14}\,\mathrm{cm^{-3}}$	$900\,\mathrm{K}$	mobility increases exponentially with rising temperature between 500 K and 900 K	
μ	$4\,\mathrm{cm^2/Vs}$			

dielectric constants:

$\varepsilon(0)$	7.1	$300\,\mathrm{K}$	infrared reflectivity	67G
$\varepsilon(\infty)$	4.5			

refractive index:

n	2.117	$300\,\mathrm{K}$, $\lambda = 0.589\,\mu\mathrm{m}$		67G

2.1.B Hexagonal boron nitride

Electronic properties

band structure: Fig. 4, Brillouin zone: Fig. 5.

   A recent band structure calculation taking into account interlayer interaction proposes an indirect gap of 3.9 eV between a valence band maximum at H and a conduction band minimum at M as well as additional interlayer conduction bands with minimum at the zone center [85C].

energy gaps (in eV):

E_g	5.2 (2)	RT	reflectance, Fig. 4	84H
	3.2 ··· 5.8		range of experimental data discussed in [84H]	
$E_{g,th}$	7.1 (1)		temperature dependence of resistivity	82C

Lattice properties

lattice parameters:

a	6.6612 Å	$297\,\mathrm{K}$		66L
c	2.5040 Å			

decomposition temperature:

T_{dec}	2600 (100) K			65J

density:

d	$2.18\,\mathrm{g\,cm^{-3}}$			

phonon dispersion relations: Fig. 6.

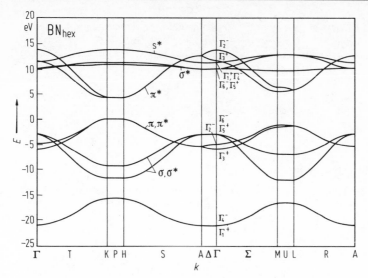

Fig. 4. BN, hexagonal. Band structure calculated with the tight binding method [84R].

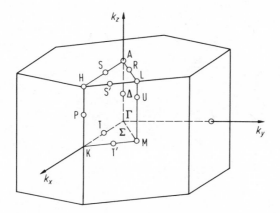

Fig. 5. Brillouin zone of the hexagonal lattice.

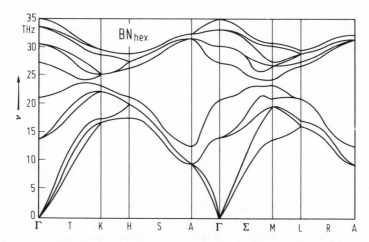

Fig. 6. BN, hexagonal. Phonon dispersion relations calculated with the rigid ion model [83S2].

Physical property	Numerical value	Experimental conditions	Experimental method, remarks	Ref.
phonon wavenumbers (in cm^{-1}):				
$\bar{v}$	49	E_{2g}	zone center Raman mode	84H
	770	A_{2u}	infrared active mode	
	1367	E_{2g}	zone center Raman mode	
	1383	E_{1u}	infrared active mode	

Transport and optical properties

Figure 7 shows the temperature dependence of the resistivity at high temperatures and a comparison with earlier literature data.

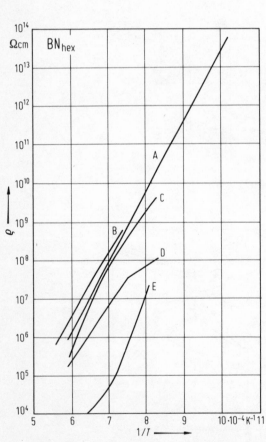

Fig. 7. BN, hexagonal. Resistivity vs. temperature above 700 °C. A: from bulk resistance of a disk measured in c direction. B···E: earlier literature data for comparison [82C].

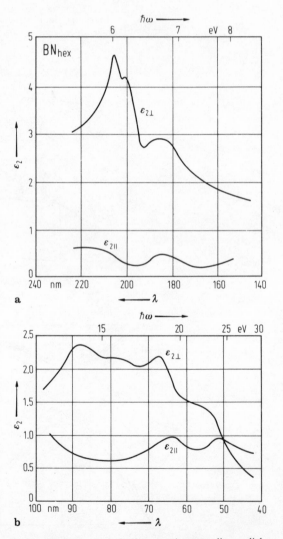

Fig. 8. BN, hexagonal. Ordinary and extraordinary dielectric functions ε_2 vs. wavelength and photon energy in the range 5···9 eV (a) and 13···30 eV (b) [81M].

Physical property	Numerical value	Experimental conditions	Experimental method, remarks	Ref.
refractive index:				
n	1.65	BN-film		81T
	1.65		perpendicular to c axis	83I
	2.13		parallel to c axis	
dielectric constants:				
$\varepsilon(0)$	5.06	300K	parallel to c axis	66G
	6.85		perpendicular to c axis	
$\varepsilon(\infty)$	4.10		parallel to c axis	
	4.95		perpendicular to c axis see also Fig. 8	

References for 2.1

57W   Wentorf, R.H.: J. Chem. Phys. **26** (1957) 956.
62P   Philipp, H.R., Taft, E.A.: Phys. Rev. **127** (1982) 159.
63S   Steigmeier, E.F.: Appl. Phys. Lett. **3** (1963) 6.
65J   Janaf Thermochemical Tables, US Dept. of Commerce NBS PB 16384 **(1965)**.
66G   Geick, R., Perry, C.H., Rupprecht, G.: Phys. Rev. **146** (1966) 543.
66L   Lynch, R.W., Drickamer, H.G.: J. Chem. Phys. **44** (1966) 181.
67G   Gielisse, P.J., Mitra, S.S., Plendl, J.N., Griffis, R.D., Mansur, L.C., Marshall, R., Pascoe, E.A.: Phys. Rev. **155** (1967) 1039.
74C   Chrenko, R.M.: Solid State Commun. **14** (1974) 511.
74S   Soma, T., Sawaoka, S., Saito, S.: Mater. Res. Bull. **9** (1974) 755.
75H   Halperin, A., Katzir, A.: J. Phys. Chem. Solids **36** (1975) 89.
76B   Bam, I.S., Davidenko, V.M., Sidorov, V.G., Fel'dgun, L.I., Skagalov, M.D., Shalabutov, Y.K.: Sov. Phys. Semicond. (English Transl.) **10** (1976) 331; Fiz. Tekh. Poluprov. **10** (1976) 554.
81M   Mamy, R., Thomas, J., Jezequel, G., Lemonnier, J.C.: J. Phys. (Paris) Lett. **42** (1981) L-473.
81T   Takahashi, T., Itoh, H., Kuroda, M.: J. Cryst. Growth **53** (1981) 418.
82C   Carpenter, L.G., Kirby, P.J.: J. Phys. **D15** (1982) 1143.
83I   Ishii, T., Sato, T.: J. Cryst. Growth **61** (1983) 689.
83S1   Sanjurjo, J.A., López-Cruz, E., Vogl, P., Cardona, M.: Phys. Rev. **B28** (1983) 4579.
83S2   Sokolovskii, T.D.: Phys. Status Solidi (b) **118** (1983) 493.
84H   Hoffmann, D.M., Doll, G.L., Eklund, P.C.: Phys. Rev. **B30** (1984) 6051.
84P   Prasad, C., Dubey, J.D.: Phys. Status Solidi (b) **125** (1984) 625.
84R   Robertson, J.: Phys. Rev. **B29** (1984) 2131.
85C   Catellani, A., Posternak, M., Baldereschi, A., Jansen, H.J.F., Freeman, A.J.: Phys. Rev. **B32** (1985) 6997.
85H   Huang, M., Ching, W.Y.: J. Phys. Chem. Solids **46** (1985) 977.

2.2 Boron phosphide (BP)

Electronic properties

band structure: Fig. 1 (Brillouin zone: Fig. 2 of section 1.1)

BP seems to be an indirect gap semiconductor. The maximum of the valence band is situated at the Γ point of the Brillouin zone. The conduction band minima are according to [72H] at the X points. According to [85H] minima at Γ, L and along the Δ axes occur within an energy range of 0.04 eV.

energy gap (in eV):

E_g	2.4	RT	analysis of diffuse reflection coefficient	83K
	2.2	Γ–X	calculated, Fig. 1	72H
	1.98	Γ–L	calculated, followed by minima along Δ (1.99 eV) and at Γ (2.02 eV)	85H

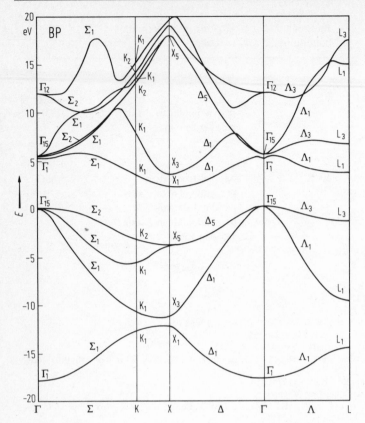

Fig. 1. BP. Band structure calculated by a non-local empirical pseudo-potential method [72H].

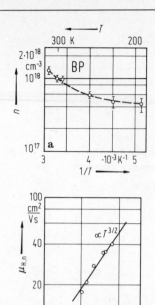

Fig. 2. BP. Electrical conductivity vs. temperature for five n-type samples [80Y].

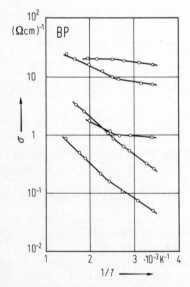

Fig. 3. BP. Electron concentration (a) and electron Hall mobility (b) vs. temperature for a single crystal [77K].

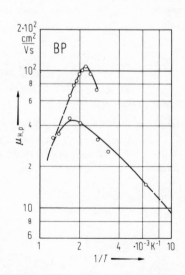

Fig. 4. BP. Hole Hall mobility vs. reciprocal temperature of two p-type single crystals with $p = 1 \cdots 5 \cdot 10^{18}$ cm^{-3} [60S].

Physical property	Numerical value	Experimental conditions	Experimental method, remarks	Ref.

Lattice properties

structure: zincblende, space group T_d^2-$F\bar{4}3m$

lattice parameter:

a	4.5383(4) Å	297 K		75S

linear thermal expansion coefficient:

α	$3.65 \cdot 10^{-6}$ K^{-1}	400 K		75S

melting point: BP decomposes at 1400 K

phonon wavenumbers:

$\bar{\nu}_{LO}$	828.9(6) cm^{-1}	RT	Raman scattering	83S
$\bar{\nu}_{TO}$	799(1) cm^{-1}			

second order elastic moduli (in 10^{12} dyn cm^{-2}):

c_{11}	5.15(1)	RT	Brillouin scattering	84W
c_{12}	1.0(1)			
c_{44}	1.60(5)			

Transport and optical properties

BP is extrinsic at RT, the transport limited by impurity scattering. Fig. 2 shows typical conductivity vs. temperature curves.

mobilities (in cm^2/Vs):

μ_n	30···40	$T = 300$ K	single crystals, $n = 10^{18}$ cm^{-3}	75I, 77K
	70···120		epitaxial films on Si, $n = 6 \cdot 10^{18} \cdots 2 \cdot 10^{21}$ cm^{-3} temperature dependence of electron mobility: Fig. 3	74S
μ_p	500	$T = 300$ K	single crystal, $p = 10^{18}$ cm^{-3}	64W
	285···350		epitaxial layers on Si, $p = 5 \cdots 8 \cdot 10^{19}$ cm^{-3}; temperature dependence of hole mobility: Fig. 4	74S

refractive index:

n	3.34(5)	$\lambda = 454.5$ nm, RT	Brewster angle method	84W
	3.34(5)	458 nm		
	3.32(5)	488 nm		
	3.30(5)	496 nm		
	3.26(5)	514.5 nm		
	3.00(5)	632.8 nm		
	3.1	589.3 nm	reflectance	76T

dielectric constant:

$\varepsilon(0)$	11	300 K	Schottky barrier reflectance	76T

References for 2.2

60S   Stone, B., Hill, D.: Phys. Rev. Lett. **4** (1960) 282.
64W   Wang, C.C., Cardona, M., Fischer, A.G.: RCA Review **25** (1964) 159.
71I   Iwami, M., Fujita, N., Kawabe, K.: Jpn. J. Appl. Phys. **10** (1971) 1746.
72H   Hemstreet, L.A., Fong, C.Y.: Phys. Rev. **B6** (1972) 1464.
74S   Sohno, K., Takigawa, M., Nakada, T.: J. Cryst. Growth **24/25** (1974) 193.
75S   Slack, G.A., Bartram, S.F.: J. Appl. Phys. **46** (1975) 89.
76T   Takenaka, T., Takigawa, M., Sohno, K.: Jpn. J. Appl. Phys. **15** (1976) 2021.
77K   Kato, N., Kamura, W., Iwami, M., Kawabe, K.: Jpn. J. Appl. Phys. **16** (1977) 1623.
80Y   Yugo, S., Kimura, T.: Phys. Status Solidi (a) **59** (1980) 363.
83K   Kurbatov, G.A., Sidorin, V.K., Sidorin, K.K., Sheludchenko, A.M.: Sov. Phys. Semicond. (English Transl.) **17** (1983)
       746; Fiz. Tekh. Poluprovodn. **17** (1983) 1180.
83S   Sanjurjo, J.A., López-Cruz, E., Vogl, P., Cardona, M.: Phys. Rev. **B28** (1983) 4579.
84W   Wettling, W., Windscheif, J.: Solid State Commun. **50** (1984) 33.
85H   Huang, M., Ching, W.Y.: J. Phys. Chem. Solids **46** (1985) 977.

Physical property	Numerical value	Experimental conditions	Experimental method, remarks	Ref.

2.3 Boron arsenide (BAs)

In the presence of As vapor the phase is stable up to 920 °C [65E], then decomposes into an orthorhombic
subarsenide (B_6As).

band structure: Fig. 1 (Brillouin zone, see Fig. 2 of section 1.1). The calculated band structure shows an indirect
(Γ–X) gap of a few eV. Experiments on this layers [74C] can be interpreted by a very small indirect gap (0.67 eV)
or more probably by a direct gap of 1.46 eV. This latter result is consistent with earlier measurements [58P].

There is almost no further information about semiconducting properties of BAs.

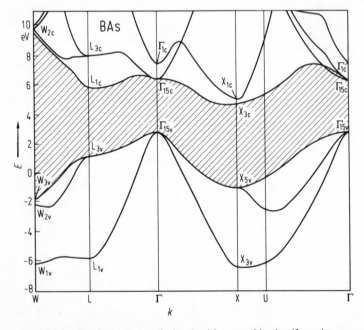

Fig. 1. BAs. Band structure calculated with a combined self-consistent
OPW and pseudopotential method [70S].

Physical property	Numerical value	Experimental conditions	Experimental method, remarks	Ref.
lattice parameter:				
a	4.777 Å			77M
density:				
d	5.22 g cm^{-3}			77M

References for 2.3

58P Perri, A., LaPlaca, S., Post, B.: Acta Crystallogr. **11** (1958) 310.
65E Elliott, R.P.: Constitution of binary alloys, first suppl., McGraw Hill, New York 1965.
70S Stuckel, D.J.: Phys. Rev. **B1** (1970) 3458.
77M Merrill, L.: J. Phys. Chem. Ref. Data **6** (1977) 1205.

2.4 Aluminum nitride (AlN)

Electronic properties

band structure: Fig. 1 (Brillouin zone: Fig. 5 of section 2.1)

AlN is a semiconductor with a large direct gap. Since it crystallizes in the wurtzite lattice the band structure differs from that of the most other III–V compounds.

energies of symmetry points of the band structure (relative to the top of the valence band) (in eV):

$E(\Gamma_{1v})$	−18.40 (−14.43)	calculated, see Fig. 1	83K
$E(\Gamma_{3v})$	−7.10 (−4.68)	(data in brackets: [85H])	
$E(\Gamma_{5v})$	−1.22 (−0.60)		
$E(\Gamma_{6v}, \Gamma_{1'v})$	0.00		
$E(\Gamma_{1c})$	6.2 (6.25)		
$E(\Gamma_{3c})$	8.92 (9.38)		
$E(\Gamma_{6c}, \Gamma_{1'c})$	13.0		
$E(L_{1'3'v})$	−7.52 (−4.26)		
$E(L_{24v})$	−1.97 (−1.06)		
$E(L_{13v})$	−1.87 (−1.03)		
$E(L_{13c})$	9.99 (9.40)		
$E(L_{1'3'c})$	13.53		

energy gap (in eV):

$E_{g,dir}$	6.2	300 K	absorption (excitonic contribution near direct edge)	79Y
	6.23	77 K		
	6.28	300 K	from excitonic edge assuming exciton binding energy of 75 meV	80R

From the dichroism of the absorption edge follows that the $\Gamma_{1'}$ state (see Fig. 1) lies slightly higher than the Γ_6 state (transition $E \parallel c$ ($\Gamma_{1'v} - \Gamma_{1c}$) at lower energy than transition $E \perp c$ ($\Gamma_{6v} - \Gamma_{1c}$)), both states being split by crystal field interaction [79Y].

Lattice properties

AlN crystallizes at normal pressure in the wurtzite structure (space group $C_{6v}^4 - P6_3mc$). A phase transition at 21 (1) GPa (tentatively to a NaCl phase) has been measured by shock compression [81K].

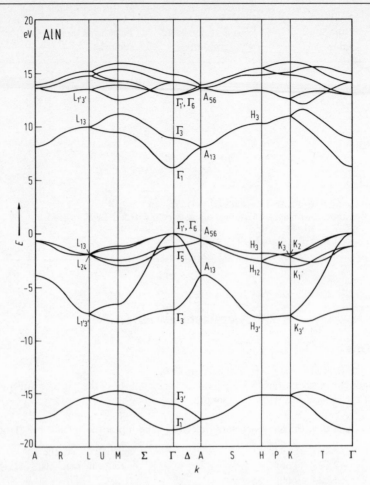

Fig. 1. AlN. Band structure calculated with a semi-empirical tight binding method [83K].

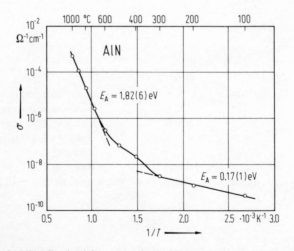

Fig. 2. AlN. Conductivity vs. reciprocal temperature for hot-pressed material [76F]. E_A: activation energy for conductivity.

Physical property	Numerical value	Experimental conditions	Experimental method, remarks	Ref.
lattice parameters:				
a	3.11(1) Å	RT	X-ray diffraction on ultrafine powder	82I
c	4.98(1) Å			
linear thermal expansion coefficient:				
$\alpha_\perp$	$5.27 \cdot 10^{-6}\,\mathrm{K}^{-1}$	$T = 20 \cdots 800\,°C$	X-ray, epitaxial layers	74S
$\alpha_\parallel$	$4.15 \cdot 10^{-6}\,\mathrm{K}^{-1}$			
melting point:				
T_m	3273 K			70M
density:				
d	$3.255\,\mathrm{g\,cm}^{-3}$		X-ray	73S
phonon wavenumbers (in cm^{-1}):				
$\bar{v}_{LO}(E_1)$	895(2)	RT	Raman scattering	83S
$\bar{v}_{TO}(E_1)$	671.6(8)			
$\bar{v}_{LO}(A_1)$	888(2)			
$\bar{v}_{TO}(A_1)$	659.3(6)			
$\bar{v}(E_2)$	303		Raman scattering, tentative assignment	84C
$\bar{v}(E_2)$	426			
$\bar{v}_{TO}(A_1)$	514			
$\bar{v}_{TO}(E_1)$	614			
$\bar{v}_{LO}(A_1)$	663			
$\bar{v}_{LO}(E_1)$	821			

Transport and optical properties

Owing to the large energy gap, transport is always extrinsic.

conductivity (in $\Omega^{-1}\mathrm{cm}^{-1}$):				
σ	$10^{-3} \cdots 10^{-5}$	$T = 290\,K$	doped (Al_2OC) single p-type crystals (blue)	65E
	$10^{-11} \cdots 10^{-13}$	300 K	undoped single crystals (colorless or pale yellow) see also Fig. 2	
hole mobility:				
μ_p	$14\,\mathrm{cm}^2/\mathrm{Vs}$	$T = 290\,K$	doped single crystal, the authors point out that this result must be viewed with some caution	65E
dielectric constants:				
$\varepsilon(0)$	9.14	300 K	reflectivity	67C
$\varepsilon(\infty)$	4.84			

References for 2.4

65E Edwards, J., Kawabe, K., Stevens, G., Tredgold, R.H.: Solid State Commun. 3 (1965) 99.
67C Collins, A.T., Lightowlers, E.C., Dean, P.J.: Phys. Rev. 158 (1967) 833.

70M   MacChesney, J.B., Bridenbaugh, P.M., O'Connor, P.B.: Mater. Res. Bull. **5** (1970) 783.
73S   Slack, G.A.: J. Phys. Chem. Solids **34** (1973) 321.
74S   Sirota, N.N., Golodushko, V.Z.: Tezisy Dokl., Vses Konf. Khi., Svyazi Poluprovdn. Polumetallakh 5th (1974) 98.
76F   Francis, R.W., Worrell, W.L.: J. Electrochem. Soc. **123** (1976) 430.
79Y   Yamashita, H., Fukui, K., Misawa, S., Yoshida, S.: J. Appl. Phys. **50** (1979) 896.
80R   Roskovcova, L., Pastrnak, J.: Czech. J. Phys. **B30** (1980) 586.
82I   Iwama, S., Hayakawa, K., Arizumi, T.: J. Cryst. Growth **56** (1982) 265.
83K   Kobayashi, A., Sankey, O.F., Volz, S.M., Dow, J.D.: Phys. Rev. **B28** (1983) 935.
83S   Sanjurjo, J.A., López-Cruz, E., Vogl, P., Cardona, M.: Phys. Rev. **B28** (1983) 4579.
84C   Carlone, C., Lakin, K.M., Shanks, H.R.: J. Appl. Phys. **55** (1984) 4010.
85H   Huang, M.Z., Ching, W.Y.: J. Phys. Chem. Solids **46** (1985) 977.

Physical property	Numerical value	Experimental conditions	Experimental method, remarks	Ref.

2.5 Aluminum phosphide (AlP)

Electronic properties

band structure: Fig. 1 (Brillouin zone: see Fig. 2 of section 1.1)

AlP is an indirect gap semiconductor, the minima of the conduction bands are situated at the X points of the Brillouin zone (no camel's back structure! [85K2]). The top of the valence band has the structure common to all zincblende semiconductors.

energies of symmetry points of the band structure (relative to the top of the valence band) (in eV):

$E(\Gamma_{1v})$	-11.82	calculated, see Fig. 1	85H
$E(\Gamma_{15v})$	0.00		
$E(\Gamma_{1c})$	3.74		
$E(\Gamma_{15c})$	5.09		
$E(X_{5v})$	-2.27		
$E(X_{1c})$	2.51		
$E(X_{3c})$	4.30		
$E(L_{3v})$	-0.80		
$E(L_{1c})$	3.57		

energy gaps (in eV):

$E_{g,ind}$ $(\Gamma_{15v} - X_{1c})$	2.505(10)	4K	excitonic gap, photo-luminescence; temperature dependence, see Fig. 2	73M
$E_{g,dir}$ $(\Gamma_{15v} - \Gamma_{1c})$	3.63(2) 3.62(2)	4 K 300 K	excitonic gap, photo-luminescence	73M

effective masses (in units of m_0):

$m_{n\parallel}$	3.67		calculated from band structure of Fig. 1	85H
$m_{n\perp}$	0.212			
$m_{p,h}$	0.513	$\parallel [100]$		
	1.372	$\parallel [111]$		
$m_{p,l}$	0.211	$\parallel [100]$		
	0.145	$\parallel [111]$		

Lattice properties

structure:

AlP I	space group $T_d^2 - F\bar{4}3m$ (zincblende lattice)	normal pressure phase	
AlP II	NaCl structure	high pressure phase	

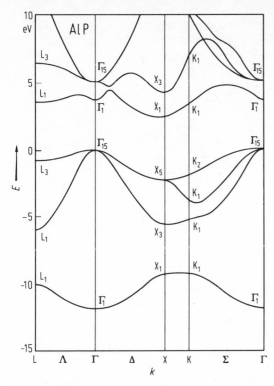

Fig. 1. AlP. Band structure calculated with an orthogonal-ized LCAO method [85H].

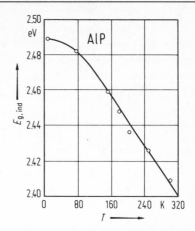

Fig. 2. AlP. Indirect energy gap vs. temperature [70M].

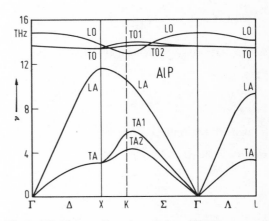

Fig. 3. AlP. Phonon dispersion relations [85K1].

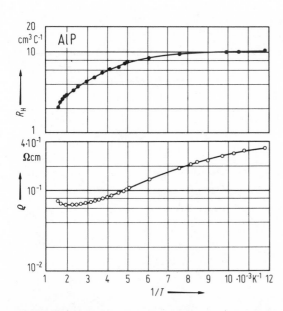

Fig. 4. AlP. Hall coefficient and resistivity vs. reciprocal temperature for an n-type sample [67R].

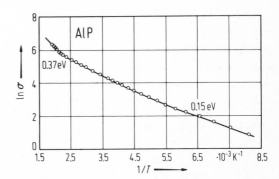

Fig. 5. AlP. Natural logarithm of conductivity vs. reciprocal temperature in an undoped sample [60G]. σ in $\Omega^{-1}\,cm^{-1}$. Activation energies are also shown.

Physical property	Numerical value	Experimental conditions	Experimental method, remarks	Ref.
transition pressure (in kbar):				
p_{tr}	140···170			78Y
lattice parameter:				
a	5.4635(4) Å	25 °C	epitaxial film on GaP	82B
density:				
d	2.40(1) g cm^{-3}			68C
melting point:				
T_m	2823 K			75O
phonon dispersion relations: Fig. 3.				
phonon wavenumbers:				
$\bar{v}_{LO}(\Gamma)$	501.0(2) cm^{-1}	300 K	Raman spectroscopy	70O
$\bar{v}_{TO}(\Gamma)$	439.4(2) cm^{-1}			
second order elastic moduli (in 10^{11} dyn cm^{-2}):				
c_{11}	18.83···14.59		theoretical estimates	85K1
c_{12}	6.71···8.44			
c_{44}	3.69···4.24			

Transport and optical properties

electron mobility:

μ_n	60 cm^2/Vs	$T = 300$ K	thin films	66R
	10···80 cm^2/Vs	298 K	single crystals, $n = 1.3 \cdot 10^{18}$ ··· $5.5 \cdot 10^{19}$ cm^{-3}	67R

The lattice mobility of holes has been estimated to 450 cm^2/Vs [75W]. See Figs. 4 and 5 for further data on transport properties.

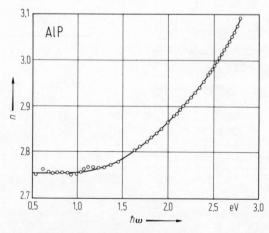

Fig. 6. AlP. Refractive index vs. photon energy at RT [70M].

Physical property	Numerical value	Experimental conditions	Experimental method, remarks	Ref.
thermal conductivity:				
κ	$0.9\,\mathrm{W\,cm^{-1}\,K^{-1}}$	$T = 300\,\mathrm{K}$		63S
dielectric constants:				
$\varepsilon(0)$	9.8		from refractive index	70M
$\varepsilon(0)/\varepsilon(\infty)$	1.3		from phonon frequencies	

refractive index: Fig. 6.

Impurities and defects

impurity levels: due to the great instability of AlP in moisture little work on defects has been reported. p-type conduction was found to arise from two acceptor species at $E_v + 0.15\,\mathrm{eV}$ and $E_v + 0.37\,\mathrm{eV}$ [60G]. These defects also seem to be involved in photoluminescence and optical absorption at energies below the band gap.

References for 2.5

60G   Grimmeiss, H.G., Kischio, W., Rabenau, A.: Phys. Chem. Solids **16** (1960) 302.
63S   Steigmeier, E.F.: Appl. Phys. Lett. **3** (1963) 6.
66R   Reid, F.J., Miller, S.E., Goering, H.L.: J. Electrochem. Soc. **113** (1966) 467.
67R   Reid, F.J., e.a.: Batelle Memoral. Inst. Ohio, High-Temperature Material Study, NASA CR-86021, June 1967, Contract No. NAS 12-107 N68-14557.
68C   Caveney, R.J.: Philos. Mag. **17** (1968) 943.
70M   Monemar, B.: Solid State Commun. **8** (1970) 1295.
70O   Onton, A.: Proc. 10th Int. Conf. Phys. Semicond., Cambridge/Mass. 1970, USAC, Oak Ridge **1970**.
71L   Lawaetz, P.: Phys. Rev. **B4** (1971) 3460.
73M   Monemar, B.: Phys. Rev. **B8** (1973) 5711.
75O   Osamura, K., Murakami, Y.: J. Phys. Chem. Solids **36** (1975) 1354.
75W   Wiley, J.D.: in "Semiconductors and Semimetals", Vol. 10, R.K. Willardson, A.C. Beer eds., Academic Press, New York **1975**.
78Y   Yu, S.C., Spain, I.L., Skelton, E.F.: Solid State Commun. **25** (1978) 49.
82B   Bessolov, V.N., Konnikov, S.G., Umanskii, V.I. Yakovlev, Yu.P.: Sov. Phys. Solid State (English Transl.) **24** (1982) 875; Fiz. Tverd. Tela **24** (1982) 1528.
85H   Huang, M., Ching, W.Y.: J. Phys. Chem. Solids **46** (1985) 977.
85K1   Kagaya, H.-M., Soma, T.: Phys. Status Solidi (b) **127** (1985) 89.
85K2   Kopylov, A.A.: Solid State Commun. **56** (1985) 1.

2.6 Aluminum arsenide (AlAs)

Electronic properties

band structure: Fig. 1 (Brillouin zone: see section 1.1, Fig. 2)

AlAs has a band structure similar to those of AlP or GaP. The minima of the conduction band are situated near the X points of the Brillouin zone.

energies of symmetry points of the band structure (relative to the top of the valence band) (in eV):

$E(\Gamma_{1v})$	$-11.95\,(-11.87)$	calculated: [85H], Fig. 1,
$E(\Gamma_{15v})$	0.00	(in brackets: [80C])
$E(\Gamma_{1c})$	2.79 (2.81)	
$E(\Gamma_{15c})$	4.48 (4.21)	
$E(X_{1v})$	$-9.63\,(-9.80)$	
$E(X_{3v})$	$-5.69\,(-5.52)$	
$E(X_{5v})$	$-2.38\,(-2.32)$	
$E(X_{1c})$	2.37 (2.21)	

Physical property	Numerical value	Experimental conditions	Experimental method, remarks	Ref.
$E(X_{3c})$	3.84 (2.89)			
$E(L_{1v})$	-5.95 (-6.41)			
$E(L_{3v})$	-0.88 (-0.97)			
$E(L_{1c})$	2.81 (2.48)			
$E(L_{3c})$	5.86 (4.87)			

energy gaps (in eV):

$E_{g,ind}$	2.229 (1)	4 K	excitonic gap, photoluminescence	73M
$(\Gamma_{15v} - X_{1c})$	2.223 (1)	77 K		
	2.153 (2)	300 K		
$E_{g,ind}$	2.363	295 K	transport	80L
$(\Gamma_{15v} - L_{1c})$				
$E_{g,dir}$	3.13 (1)	4 K	excitonic gap, photoluminescence	73M
$(\Gamma_{15v} - \Gamma_{1c})$	3.03 (1)	300 K		

exciton binding energy:

E_b	0.0258 eV		calculation including camel's back structure	81B

For the temperature dependence of the indirect and direct energy gap, see Fig. 2.

critical point energies and spin-orbit splitting energies (in eV):

E_0	3.02	300 K	direct gap, see also tables above	
$E_0 + \Delta_0$	3.32		data from [70O], assignment	
$E_1(1)$	3.83		corrected by [71B] (from a	
$E_1(2)$	3.96		compilation in [85A])	
$E_1(1) + \Delta_1$	4.03			
$E_1(2) + \Delta_1$	4.16			
E_0'	4.54			
$E_0' + \Delta_0'$	4.69			
E_2	4.89			

camel's back structure of conduction band edge:
A camel's back structure of the bottom of the conduction band near X (see respective table and Fig. 2 of section 2.9) has been proved experimentally [81B].

Δ	411 meV		extrapolated using GaP data	85K
ΔE	0.2 meV			
k_m	0.042 $(2\pi/a)$			
m_t	$0.227 m_0$			
$m_\parallel$	$10.7 m_0$			
m_l	$1.56 m_0$		longitudinal mass far above the band minimum (polaron correction included)	

effective masses, electrons (in units of m_0):
from a compilation and discussion of literature data in [85A]

$m_{n\perp}(X)$	0.19		effective masses at X neglecting camel's back structure	
$m_{n\parallel}(X)$	1.1			
$m_{n,ds}(X)$	0.71		density of states masses obtained	
$m_{n,c}(X)$	0.26		from $m_{ds} = N^{2/3} m_\perp^{2/3} m_\parallel^{1/3}$ with	
$m_{n\perp}(L)$	0.0964		$N =$ number of eq. minima	

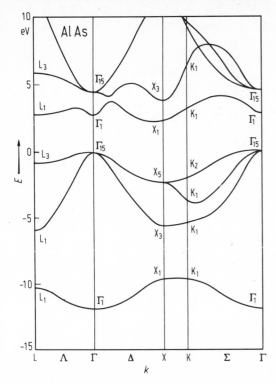

Fig. 1. AlAs. Band structure calculated with an orthogonal-ized LCAO method [85H].

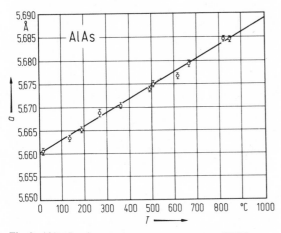

Fig. 3. AlAs. Lattice parameter vs. temperature [70E].

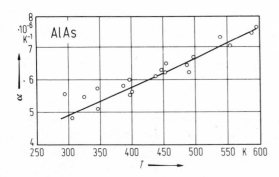

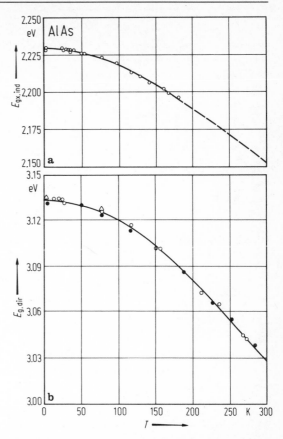

Fig. 2. AlAs. (a). Indirect excitonic gap and (b) direct gap vs. temperature. Due to a high impurity content the curve in (b) is believed to represent an energy a few meV above the exciton edge in pure AlAs [73M].

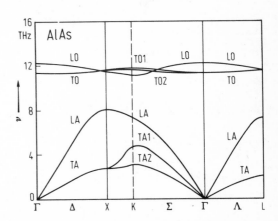

Fig. 5. AlAs. Phonon dispersion relations [83K].

Fig. 4. AlAs. Coefficient of linear thermal expansion vs. tem-perature above RT [59P].

Physical property	Numerical value	Experimental conditions	Experimental method, remarks	Ref.
$m_{n\parallel}(L)$	1.9		conductivity effective	
$m_{n,ds}(L)$	0.66		mass obtained from	
$m_{n,c}(L)$	0.14		$m_c = (2/m_\perp + 1/m_\parallel)^{-1}$	
$m_n(\Gamma)$	0.150		same values for m_{ds} and m_c	

effective masses, holes (in units of m_0):

$m_{p,h}$	0.409	$\parallel[100]$	calculated from band	85H
	1.022	$\parallel[111]$	structure data (see above)	
$m_{p,1}$	0.153	$\parallel[100]$		
	0.109	$\parallel[111]$		

valence band parameters:

A	-4.03			75W
B	-2.09			
$|C|$	4.63			

Lattice properties

structure:
AlAs crystallizes in the zincblende structure (space group $T_d^2-F\bar{4}3m$). A high-pressure phase has not been observed.

lattice parameter:

a	5.660 Å	291.15 K	for temperature dependence, see Fig. 3 [70E]	67P

coefficient of linear thermal expansion: see Fig. 4.

density:

d	3.760 g cm^{-3}		calculated from lattice parameter	85A

melting point:

T_m	2013 (20) K			64K

phonon dispersion curves: Fig. 5.

phonon wavenumbers (energies) (in cm^{-1} and meV, respectively):

$\bar{v}_{LO}(\Gamma)$	403.7 (50.09)
$\bar{v}_{TO}(\Gamma)$	361.7 (44.88)

second order elastic moduli (in 10^{11} dyn cm^{-2}):

c_{11}	12.02		calculated with an empirical	85A
c_{12}	5.70		interpolation formula using	
c_{44}	5.89		data from other III–V compounds	

Transport and optical properties

Only few reliable data exist on the transport and optical properties of AlAs. Most results are extrapolations from data obtained on the technologically more important solid solutions of the type $Al_xGa_{1-x}As$.

Electron mobilities have been reported up to 300 cm^2/Vs at RT, hole mobilities should be about 200 cm^2/Vs for pure lattice scattering [75W].

Physical property	Numerical value	Experimental conditions	Experimental method, remarks	Ref.
electron mobility:				
μ_n	$294 \cdots 75 \, \text{cm}^2/\text{Vs}$	$T = 300 \, \text{K}$	single crystal layers	71E
	$18 \, \text{cm}^2/\text{Vs}$	77 K		

For temperature dependence of resistivity, carrier concentration and Hall mobility, see Figs. 6 and 7.

dielectric constants:				
$\varepsilon(0)$	10.06 (4)			71F
$\varepsilon(\infty)$	8.16 (2)			

refractive index: Fig. 8.

Impurities and defects

binding energy of donors (in eV):

E_b	0.07	$T = 300 \, \text{K}$	Si-doped sample	70K
	0.06		Zn-doped sample	
	0.05		Mg-doped sample	

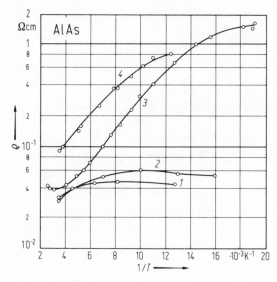

Fig. 6. AlAs. Resistivity vs. reciprocal temperature for four samples [65W].

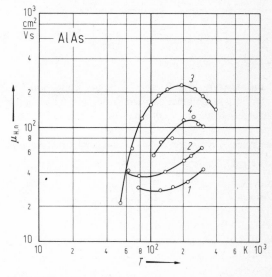

Fig. 7. AlAs. Electron Hall mobilities vs. temperature for the samples of Fig. 6 [65W].

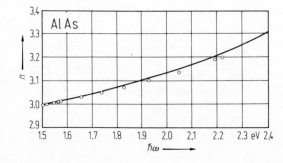

◄

Fig. 8. AlAs. Refractive index vs. photon energy for the range $1.5 \cdots 2.4 \, \text{eV}$. Data from [71F], solid curve calculated by [85C].

References for 2.6

59P   Pashintsev, I.I., Sirota, N.N.: Dokl. Akad. Nauk SSSR **3** (1959) 38.
64K   Kischio, W.: Z. Anorg. Allg. Chem. **328** (1964) 187.
65W   Whitaker, J.: Solid State Electron. **8** (1965) 649.
67P   Pearson, W.B.: A Handbook of Lattice Spacing and Structure of Metals and Alloys, Pergamon Press, Oxford-Landon **1967**.
70E   Ettenberg, M., Pfaff, R.J.: J. Appl. Phys. **41** (1970) 3926.
70K   Kressel, H., Nicoll, F.H., Ettenberg, M., Yim, W.M., Sigai, A.G.: Solid State Commun. **8** (1970) 1407.
70O   Onton, A.: Proc. 10th Int. Conf. Phys. Semicond., Cambridge/Mass. 1970, USAEC, New York **1970**, p. 107.
71B   Berninger, W.H., Rediker, R.H.: Bull. Am. Phys. Soc. **16** (1971) 305.
71E   Ettenberg, M., Sigai, A.G., Dreeben, A., Gilbert, S.L.: J. Electrochem. Soc. **119** (1971) 1355.
71F   Fern, R.E., Onton, A.: J. Appl. Phys. **42** (1971) 3499.
73M   Monemar, B.: Phys. Rev. **B8** (1973) 5711.
75W   Wiley, J.D.: in "Semiconductors and Semimetals", Vol. 10, R.K. Willardson and A.C. Beer eds., Academic Press, New York and London **1975**.
80C   Chen, A., Sher, A.: Phys. Rev. **B22** (1980) 3886.
80L   Lee, H.J., Juravel, L.Y., Woolley, J.C., SpringThorpe, A.J.: Phys. Rev. **B21** (1980) 659.
81B   Bimberg, D., Bludau, W., Linnebach, R., Bauser, E.: Solid State Commun. **37** (1981) 987.
83K   Kagaya, H.-M., Soma, T.: Solid State Commun. **48** (1983) 785.
85A   Adachi, S.: J. Appl. Phys. **58** (1985) R1.
85C   Campi, D., Papuzza, C.: J. Appl. Phys. **57** (1985) 1305.
85H   Huang, M., Ching, W.Y.: J. Phys. Chem. Solids **46** (1985) 977.
85K   Kopylov, A.A.: Solid State Commun. **56** (1985) 1.

Physical property	Numerical value	Experimental conditions	Experimental method, remarks	Ref.

2.7 Aluminum antimonide (AlSb)

Electronic properties

band structure: Fig. 1 (Brillouin zone: see section 1.1, Fig. 2)

AlSb is an indirect gap semiconductor with conduction band minima near X. The valence band shows the structure common to all zincblende semiconductors.

energies of symmetry points of the band structure (relative to the top of the valence band) (in eV):

calculated, see Fig. 1   85H

$E(\Gamma_{15v})$	0.00
$E(\Gamma_{1c})$	2.05
$E(\Gamma_{15c})$	3.50
$E(X_{5v})$	-2.31
$E(X_{1c})$	2.08
$E(X_{3c})$	3.02
$E(L_{3v})$	-0.90
$E(L_{1c})$	1.94

energy gaps (in eV):
(single group symmetry notation; taking the spin-orbit splitting of the valence band into account, the Γ_{15v}-band corresponds to the upper spin-orbit split Γ_{8v}-band)

$E_{g,ind}$	1.615(3)	300 K	excitonic gap (see Fig. 2),	83A
$(\Gamma_{15v} - \Delta_{1c})$	1.686(1)	27 K	modulation spectroscopy	
$E_{g,ind}$	2.211	295 K	electroreflectance, see Fig. 3	
$(\Gamma_{15v} - L_{1c})$	2.327	35 K	for temperature dependence (critical point E_L)	
$E_{g,dir}$	2.300	295 K	electroreflectance, see Fig. 3	82J
$(\Gamma_{15v} - \Gamma_{1c})$	2.384	25 K	for temperature dependence (critical point E_0)	

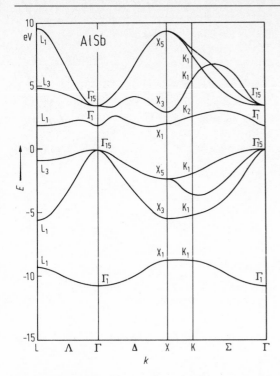

Fig. 1. AlSb. Band structure calculated with an orthogonalized LCAO method [85H].

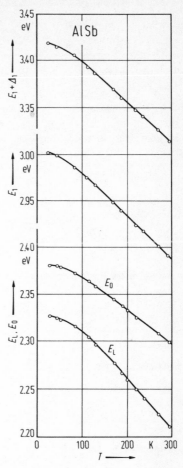

Fig. 3. AlSb. Critical point energies vs. temperature [82J].

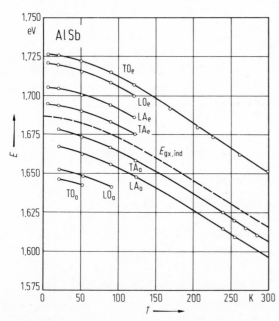

Fig. 2. AlSb. Indirect excitonic energy gap and energies of phonon assisted indirect transitions vs. temperature [83A]. Subscripts e and a stand for transitions involving emission and absotption of phonons.

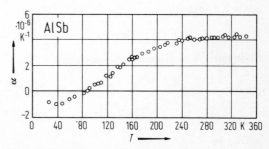

Fig. 4. AlSb. Coefficient of linear thermal expansion vs. temperature [63N].

Physical property	Numerical value	Experimental conditions	Experimental method, remarks	Ref.
spin-orbit splitting energies:				
$\Delta_0(\Gamma_{8v} - \Gamma_{7v})$	673 meV	295 K	splitting of Γ_{15v}	82J
$\Delta_1(L_{4,5v} - L_{6v})$	426 meV	295 K	splitting of L_{3v}	

camel's back structure of conduction band edge: (see respective table and Fig. 2 of section 2.9 for meaning of symbols)

Δ	261 meV		estimates using $k \cdot p$ theory and GaP data	85K2
ΔE	7.4 meV			85K2
k_m	$0.101\,(2\pi/a)$			
m_t	$0.259\,m_0$			
$m_\parallel$	$1.8\,m_0$			

effective masses, holes (in units of m_0):

$m_{p,h}$	0.336	$\parallel [100]$	theoretical estimates	85H
	0.872	$\parallel [111]$		
$m_{p,1}$	0.123	$\parallel [100]$		
	0.091	$\parallel [111]$		

valence band parameters:

A	-4.12		calculation using $k \cdot p$ theory	75W
B	-2.09			
$|C|$	4.71			

Lattice properties

structure:

AlSb I	space group T_d^2–$F\bar{4}3m$ (zincblende structure)		stable at normal pressure	
AlSb II	orthorhombic distortion of NaCl structure, space group D_{2h}^{23}–Fmmm?		high-pressure phase	82B

transition pressure:

p_{tr}	77 (5) kbar		beginning of phase transition	82B

lattice parameter:

a	6.1355 (1) Å	291.5 K	powder, X-ray measurement	58G

linear thermal expansion coefficient: Fig. 4.

density:

d	4.26 g cm^{-3}	293.15 K	slightly temperature dependent	69G

melting point:

T_m	1338 K			75O

phonon dispersion relations: Fig. 5.

Physical property	Numerical value	Experimental conditions	Experimental method, remarks	Ref.
phonon wavenumbers (in cm^{-1}):				
$\bar{v}_{LO}(\Gamma)$	340.0(7)		first order Raman scattering	86V
$\bar{v}_{TO}(\Gamma)$	318.7(7)			

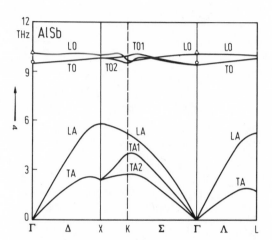

Fig. 5. AlSb. Phonon dispersion relations. Circle and triangle: experimental values for the zone center phonons [85K1].

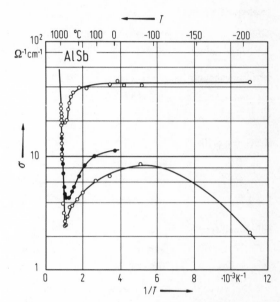

Fig. 6. AlSb. Conductivity vs. reciprocal temperature for three different polycrystalline p-type samples [53W].

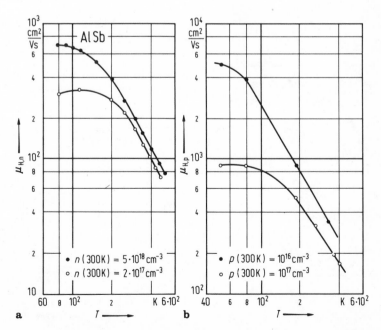

Fig. 7. AlSb. (a) Electron Hall mobility vs. temperature [66S], (b) hole Hall mobility vs. temperature [58R].

Physical property	Numerical value	Experimental conditions.	Experimental method, remarks	Ref.

second order elastic moduli (in 10^{11} dyn cm^{-2}):

c_{11}	8.769 (20)	300 K	ultrasound, values reevaluated	60B
c_{12}	4.341 (20)		by [72W] using the correct	
c_{44}	4.076 (8)		density	

Transport properties

AlSb grown without intentionally doping becomes intrinsic at about 1000 K. At high temperatures acoustic-mode scattering dominates for the holes and polar optical-mode scattering for the electrons.

intrinsic conductivity $\sigma_i = \sigma_0 \exp(-E_{g,th}/2kT)$:

σ_0	$1.12 \cdot 10^4 \,\Omega^{-1}\,\mathrm{cm}^{-1}$		see also Fig. 6.	58N
$E_{g,th}$	1.57 eV			

electron mobility:

μ_n	200 cm^2/Vs	295 K	single crystal, Te-doped	66S
	700 cm^2/Vs	77 K	$n = 4.7 \cdot 10^{16}$ cm^{-3}	

See Fig. 7a [66S] for temperature dependence ($T^{-1.8}$-dependence above 200 K).

hole mobility:

μ_p	400 cm^2/Vs	300 K	single crystal,	58R
	2000 cm^2/Vs	77 K	p (300 K) $\approx 10^{16}$ vm^{-3}	66S

See Fig. 7b [58R] for temperature dependence ($T^{-2.2}$-dependence above 100 K).

thermal conductivity: Fig. 8

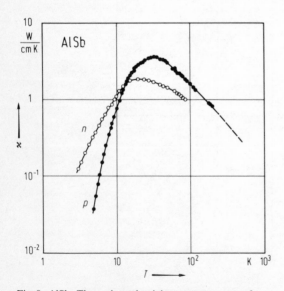

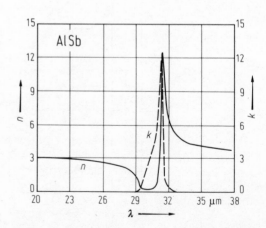

Fig. 8. AlSb. Thermal conductivity vs. temperature for an n-type and a p-type sample, low temperature range [69M].

Fig. 9. AlSb. Refractive index n and extinction coefficient k vs. wavelength in the region of lattice absorption [62T].

Physical property	Numerical value	Experimental conditions	Experimental method, remarks	Ref.

Optical properties

refractive index:

		$T\,[K]$, $\lambda\,[\mu m]$			
n	3.652	300	40	transmission and reflectance,	62T
	2.995		20	Fig. 9	
	2.080		15		
	3.100		10	transmission and reflectance	54O
	3.182		4		
	3.300		2		
	3.445		1.1		

absorption index:

k	0.0006	300,	40	transmission and reflectance,	62T
	11.73		31.3	Fig. 9	
	0.001		20		

dielectric constants:

$\varepsilon(0)$	12.04	300 K	ir reflectance, oscillator fit	62H
$\varepsilon(\infty)$	10.24			

Impurities and defects

energy levels of donors and acceptors (in meV):
No detailed analysis of donor states has yet been reported since the camel's back structure of the conduction band minimum leads to a strongly non-parabolic longitudinal effective mass.

$E_c - E_d$	147	Se-doped, photoexcitation spectrum	68A
	160	Hall data (thermal activation energy)	60T
	68	Te-doped, Hall data	
$E_a - E_v$	42.5	effective mass estimate	
	40.4	unidentified species, from photoexcitation	68A
	41.9	spectra under uniaxial stress	
	102.7		

References for 2.7

53W Welker, H.: Z. Naturforsch, **8a** (1953) 248.
54O Oswald, F., Schade, R.: Z. Naturforsch. **9a** (1954) 611.
58G Giesecke, G., Pfister, H.: Acta Crystallogr. **11** (1958) 369.
58N Nasledov, D.N., Slobodchikov, S.V.: Sov. Phys. Techn. Phys. **3** (1958) 669.
58R Reid, F.J., Willardson, R.K.: J. Electron. Control **5** (1958) 54.
60B Bolef, D.I., Menes, M.: J. Appl. Phys. **31** (1960) 1426.
60T Turner, W.J., Reese, W.E.: Phys. Rev. **117** (1960) 1003.
62H Hass, M., Henvis, B.W.: J. Phys. Chem. Solids **23** (1962) 1099.
62T Turner, W.J., Reese, W.E.: Phys. Rev. **127** (1962) 126.
63N Novikova, S.I., Abrikhosov, N.Kh.: Sov. Phys. Solid State (English Transl.) **5** (1963) 1558; Fiz. Tverd. Tela **5** (1963) 2138.
66S Stirn, R.J., Becker, W.M.: Phys. Rev. **141** (1966) 621, **148** (1966) 90,7.
68A Ahlburn, B.T., Ramdas, A.K.: Phys. Rev. **167** (1968) 717.
69G Glazov, V.M., Chizhevskaya, S.N., Evgen'ev, S.B.: Zh. Fiz. Khim. **43** (1969) 373.
69M Muzhdaba, V.M.; Nashel'skii, A.Ya., Tamarin, P.V., Shalyt, S.S.: Sov. Phys. Solid State (English Transl.) **10** (1969) 2265; Fiz. Tverd. Tela **10** (1968) 2866.
71L Lawaetz, P.: Phys. Rev. **B4** (1971) 3460
72W Weil, R.: J. Appl. Phys. **43** (1972) 4271.

75O   Osamura, K., Murakami, Y.: J. Phys. Chem. Solids **36** (1975) 931.
75W   Wiley, D.J.: in "Semiconductors and Semimetals", vol. 10 Willardson, R.K., Beer, A.C. eds., Academic Press, New York 1975.
82B   Baublitz, M., Ruoff, A.L.: J. Appl. Phys. **54** (1980) 2109.
82J   Joullié, A., Girault, B., Joullié, A.M., Zien-Eddine, A.: Phys. Rev. **B25** (1982) 7830.
83A   Alibert, C., Joullié, A., Joullié, A.M., Ance, C.: Phys. Rev. **B27** (1983) 4946.
85H   Huang, M., Ching, W.Y.: J. Phys. Chem. Solids **46** (1985) 977.
85K1   Kagaya, H.M., Soma, T.: Phys. Status Solidi (b) **127** (1985) 89.
85K2   Kopylov, A.A.: Solid State Commun. **56** (1985) 1.
86V   Ves, S., Strössner, K., Cardona, M.: Solid State Commun. **57** (1986) 483.

Physical property	Numerical value	Experimental conditions	Experimental method, remarks	Ref.

2.8 Gallium nitride (GaN)

Electronic properties

band structure: Fig. 1 (Brillouin zone: Fig. 5 of section 2.1).

The band structure given in Fig. 1 differs only slightly from other spin-neglecting calculations. Introduction of spin-orbit interaction leads to a splitting of the uppermost valence band at Γ from $\Gamma_1 + \Gamma_6$ into $\Gamma_9 + \Gamma_7 + \Gamma_7$. The energy differences between these terms can be described by two parameters – the spin-orbit splitting energy Δ_{so} and the crystal field splitting energy Δ_{cr}.

energy gap (in eV):

$E_{g,dir}$	3.503(2)	$T = 1.6$ K	photoluminescence, from excitonic gap adding the exciton binding energy	74M
	3.4751(5)		A-exciton (transition from Γ_{9v})	
	3.4815(10)		B-exciton (transition from upper Γ_{7v})	
	3.493(5)		C-exciton (transition from lower Γ_{7v})	
	3.44	300 K	temperature dependence below 295 K given by: $E_g(T) - E_g(0) = -5.08 \cdot 10^{-4} T^2/(996 - T)$, see Fig. 2 ($T$ in K)	

intra valence band energies:

The energy separations between the Γ_9 state and the two Γ_7 states can be calculated from the energy separations of the A-, B-, C-excitons.

Δ_{cr}	22(2) meV		calculated from the values given above	71D
Δ_{so}	11(+5, −2) meV			

effective masses (in m_0):

m_n	0.27(6)	$T = 300$ K	Faraday rotation	74R
$m_{n\perp}$	0.20(2)		fit of reflectance spectrum	73B
$m_{n\parallel}$	0.20(6)			
m_p	0.8(2)			75P

Lattice properties

structure:

GaN crystallizes in the wurtzite structure, space group $P6_3 mc$.

lattice parameter:

a	3.160 ⋯ 3.190 Å			79L
c	5.125 ⋯ 5.190 Å			

temperature dependence of lattice parameter: Fig. 3.

Physical property	Numerical value	Experimental conditions	Experimental method, remarks	Ref.

linear thermal expansion coefficient: Fig. 4.

phonon dispersion relations: not yet determined. Nine optical branches.

phonon wavenumbers (in cm^{-1}):

$\bar{\nu}_{A1}(TO\parallel)$	533	$T = 300\,K$	Raman spectroscopy	70M
$\bar{\nu}_{E1}(TO\perp)$	559		Raman spectroscopy	72L
$\bar{\nu}_{E1}(LO\perp)$	746		Kramers-Kronig analysis of	73B
$\bar{\nu}_{A1}(LO)$	744		infrared reflectivity	

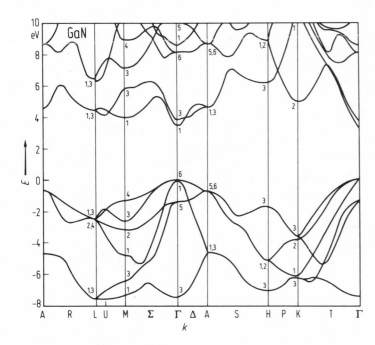

Fig. 1. GaN. Band structure calculated with an empirical pseudopotential method [74B].

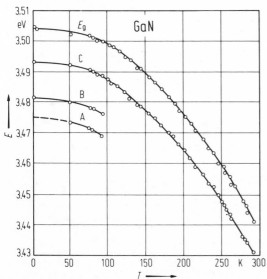

Fig. 2. GaN. Band gap energy and exciton energies vs. temperature [74M].

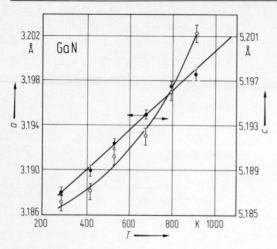

Fig. 3. GaN. Lattice parameters a and c vs. temperature for a single crystal layer [69M].

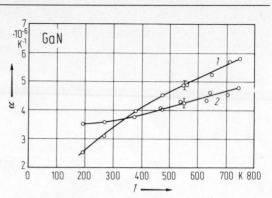

Fig. 4. GaN. Coefficient of linear thermal expansion vs. temperature; curve *1*: $\alpha_\perp$, *2*: $\alpha_\parallel$ [76S].

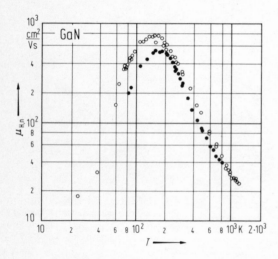

Fig. 5. GaN. Electron Hall mobility vs. temperature for two samples [73I2].

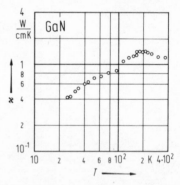

Fig. 6. GaN. Thermal conductivity along the c-axis vs. temperature [77S].

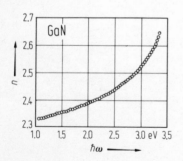

Fig. 7. GaN. Refractive index vs. photon energy at 300 K; $E \perp c$ [71E].

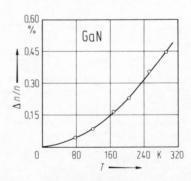

Fig. 8. GaN. Long-wavelength refractive index normalized to the 0 K value vs. temperature [71E].

Physical property	Numerical value	Experimental conditions	Experimental method, remarks	Ref.

second order elastic moduli (in 10^{11} dyn cm^{-2}):

c_{11}	29.6(1.8)		calculated from the mean square
c_{12}	13.0(1.1)		displacement of the lattice
c_{13}	15.8(0.6)		atoms measured by X-ray
c_{33}	26.7(1.8)		diffraction
c_{44}	2.41(20)		

Transport properties

GaN is an extrinsic n-type semiconductor, p-type material does not seem achievable. Above room temperature transport is predominantly determined by polar-optical scattering and at lower temperatures by impurity scattering. Crystals with n larger than $8\cdot10^{18}$ cm^{-3} show metallic conduction with no appreciable variation in n or μ_n between 10 and 300 K.

conductivity:

σ	$6\cdots12\,(\Omega\text{cm})^{-1}$	$T=300$ K	purest material, $n\approx10^{17}$ cm^{-3}, undoped layers grown by vaporphase technique on sapphire	72I, 73I1, 78C

electron mobility:

μ_n	$\leqq440$ cm^2/Vs	$T=300$ K	purest material, $n\approx10^{17}$ cm^{-3}	72I, 72I1, 78C

For temperature dependence, see Fig. 5.

thermal conductivity: Fig. 6.

Optical properties

refractive index:

n	2.29(5)	$T=300$ K (extrapolated to 0 eV), $E\perp c$	interference method (the value for $E\parallel c$ is 1.5(2)% lower at 500 nm); for energy dependence, see Fig. 7, for long wavelength value, see Fig. 8.	71E

dielectric constants:

$\varepsilon(0)$	10.4(3)	$E\parallel c$	$\varepsilon(\infty)$ assumed to be isotropic	73B
	9.5(3)	$E\perp c$		
$\varepsilon(\infty)$	5.8(4)	$T=300$ K, $E\parallel c$		70M
	5.35(20)	$E\perp c$		73B

Impurities and defects

donors

Undoped GaN is n-type with electron concentrations $n\geqq10^{17}$ cm^{-3}. The dominant donor is believed to be the nitrogen vacancy [75P]. The binding energy of this donor is subject to large controversy. Values given range from 42(1) meV [71D1] over 29(6) meV [74L] to 17 meV [79V]. A deep donor level with $E_d=110$ meV is also postulated [79V]. No detailed information exists about the identity of these donors.

acceptors

Most data have been derived from photoluminescence peaks which are caused by the recombination of free or bound electrons with bound holes.

luminescence peak energies E_{peak}.

Impurity	E_{peak} [eV]	T [K]	Remarks	Ref.
Li	2.23	77	photoluminescence	73P
Be or V_{Ga}	3.264	4.2		73I1
Be-complex or Be_{Ga}	2.36			80M1
Mg or V_{Ga}	3.264			73I1
Mg-complex	2.95			73I1
or Mg_{Ga}				80M1
Dy	3.15	77		73P
Cd or V_{Ga}	3.268	4.2		80M1
	3.263	1.6		74L,
Cd-complex or Cd_{Ga}	2.7			80M1
Zn or V_{Ga}	3.268	4.2		80M1
Zn-complex or Zn_{Ga}	2.9			74L,
(A-band)				80M1
Zn_N(?) (B-band)	2.6			80M1
Zn_N^- (?) (C-band)	2.2			
Zn_N^{2-} (?) (D-band)	1.8			

deep defect states

Undoped GaN is invariably n-type usually with a high free electron concentration ($10^{17}\cdots10^{18}$ cm^{-3}) at room temperature. Although the dominant donor has not been unambiguously identified there is a consensus that it is an intrinsic defect and most probably the nitrogen vacancy [75P]. Energies are reported ranging from 17 meV [79V] to 42 meV [71D]. The resistivity of the material can be increased dramatically by the addition of group II atoms; these elements introduce deep acceptors which compensate the native donors. The same binding energy of 225 meV is reported [74L] for the addition of Zn, Cd, Mg, and Be, however it seems unlikely that these four dissimilar species would be identical and consequently it has been proposed that the acceptor is actually V_{Ga} [80M1]. The vacancy concentration is believed to be increased by the presence of the group II element.

binding energies of acceptors

Impurity	E_b [meV]	T [K]	Remarks	Ref.
V_{Ga}	225	4.2	photoluminescence	80M2
Hg	410	78	infrared quenching of luminescence	74E
Li	750			
Zn_{Ga}(A-band)	480	4.2		
	370(40)	4.2	photoluminescence	80M2
Zn_N(B-band)	650(80)			
Zn_N^- (C-band)	1020(50)			
Zn_N^{2-} (D-band)	1420(80)			

References for 2.8

69M   Maruska, H.P., Tietjen, J.J.: Appl. Phys. Lett. **15** (1969) 327.
70M   Manchon, D.D., Barker, A.S., Dean, J.P., Zetterstrom, R.B.: Solid State Commun. **8** (1970) 1227.
71D   Dingle, R., Ilegems, M.: Solid State Commun. **9** (1971) 175.
71E   Ejder, E.: Phys. Status Solidi (a) **6** (1971) K39.
72I   Ilegems, M.: J. Cryst. Growth **13/14** (1972) 360.
72L   Lemos, V., Argüello, C.A., Leite, R.C.C.: Solid State Commun. **11** (1972) 1351.
73B   Barker, A.S., Ilegems, M.: Phys. Rev. **B7** (1973) 743.
73I1 Ilegems, M., Dingle, R.: J. Appl. Phys. **44** (1973) 4234.
73I2 Ilegems, M., Montgomery, H.C.: J. Phys. Chem. Solids **34** (1972) 885.
73P   Pankove, J.I., Duffy, M.T., Miller, E.A., Berkeyheiser, J.E.: J. Lumin. **8** (1973) 89.
74B   Bloom, S., Harbeke, G., Meier, E., Ortenburger, I.B.: Phys. Status Solidi (b) **66** (1974) 161.
74L   Lagerstedt, O., Monemar, B.: J. Appl. Phys. **45** (1974) 2266.
74M   Monemar, B.: Phys. Rev. **B10** (1974) 676.
74R   Rheinländer, B., Neumann, H.: Phys. Status Solidi (b) **64** (1974) K123.
75P   Pankove J.I., Bloom, S., Harbeke, G.: RCA Rev. **36** (1975) 163.
76S   Sheleg, A.U., Savastenko, V.A.: Vesti Akad. Nauk BSSR, Ser. Fiz. Mat. Nauk **3** (1976) 126.
77S   Sichel, E.K., Pankove, J.I.: J. Phys. Chem. Solids **38** (1977) 330.
78C   Crouch, R.K., Debnam, W.J., Fripp, A.L.: J. Mater. Sci. **13** (1978) 2358.
78S   Savastenko, V.A., Sheleg, A.U.: Phys. Status Solidi (a) **48** (1978) K135.
79L   Lagerstedt, O., Monemar, B.: Phys. Rev. **B19** (1979) 3064.
79V   Vavilov, V.S., Makarov, S.I., Chukichev, M.V., Chetverikova, I.F.: Sov. Phys. Semicond. (English Transl.) **13** (1979) 1259; Fiz. Tekh. Poluprov. **13** (1979) 2153.
80M1 Monemar, B., Lagerstedt, O., Gislason, H.P.: J. Appl. Phys. **51** (1980) 625.
80M2 Monemar, B., Gislason, H.P., Lagerstedt, O.: J. Appl. Phys. **51** (1980) 640.

Physical property	Numerical value	Experimental conditions	Experimental method, remarks	Ref.

2.9 Gallium phosphide (GaP)

Electronic properties

band structure: Fig. 1 (Brillouin zone: see section 1.1, Fig. 2)

GaP is an indirect gap semiconductor. The lowest set of *conduction bands* shows a camel's back structure (Fig. 2); the band minima are situated at the Δ-axes near the zone boundary. The *valence bands* show the usual structure characteristic for all zincblende semiconductors.

The spin-orbit splitting of the top of the valence band is negligible compared with most other energy separations in the band structure. Thus, Fig. 1 shows the band structure calculated without inclusion of the spin-orbit interaction using the single group notation for the high symmetry band states.

energies of symmetry points of the band structure (relative to the top of the valence band) (in eV):

$E(\Gamma_{1v})$	−12.99	−12.3	−13.2	symmetry symbols in single
$E(\Gamma_{15v})$	0.00			group notation
$E(\Gamma_{1c})$	2.88			first row: theoretical data
$E(\Gamma_{15c})$	5.24			according to [76C]
$E(X_{1v})$	−9.46		−9.6	second row: experimental
$E(X_{3v})$	−7.07	−6.8		data from angle resolved
$E(X_{5v})$	−2.73	−3.0	−2.7	photoemission [84S1]
$E(X_{1c})$	2.16			see also Fig. 1
$E(X_{3c})$	2.71			third row: XPS data [74L]
$E(L_{1v})$	−10.60	−10.8	−10.6	
$E(L_{2v})$	−6.84	−6.8	−6.9	
$E(L_{3v})$	−1.10	−0.9	−1.2	
$E(L_{1c})$	2.79			
$E(L_{3c})$	5.74			
$E(\Sigma_1^{min})$	−4.3	−3.8	−4.0	

Physical property	Numerical value	Experimental conditions	Experimental method, remarks	Ref.

indirect energy gap (in eV):

$E_{g,ind}$ ($\Gamma_{15v} - \Delta_{1c}$)	2.350(1)	0K, extra-polated	from exciton data obtained by wavelength modulated	78H
	2.272	300 K	transmission	

Temperature dependence: Fig. 3. The curve in Fig. 3 can be approximated by the formula $E_{g,ind}$ (eV) = 2.338 − $6.2 \cdot 10^{-4} T^2/(T + 460)$ (T in K).

$E_{gx}(1S, X_7)$	2.3284	4.2 K	excitonic gap, this value corresponds to the dip of the camel's back (see below)	78H

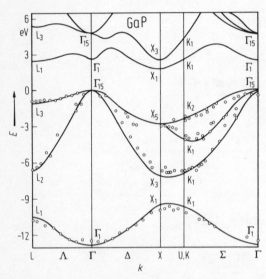

Fig. 1. GaP. Band structure calculated by a pseudopotential method neglecting spin-orbit interaction [76C]; circles: data from angle resolved photoemission [84S1].

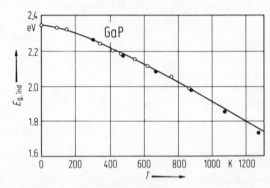

Fig. 3. GaP. Indirect energy gap vs. temperature from various authors. The solid curve was calculated by the formula given in the tables [69P].

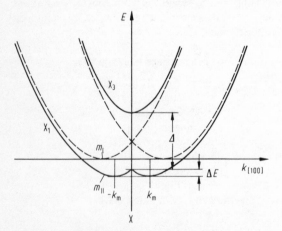

Fig. 2. GaP. Camel's back structure of the conduction band minima at the Δ-axes near the zone boundary X. Dashed curves: diamond structure, solid curves: zincblende structure. The higher band has X_3 (X_7) symmetry, the lower band X_1 (X_6) symmetry at the zone boundary.

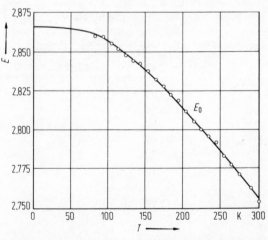

Fig. 4. GaP. Direct exciton edge E_0 and spin-orbit split edge $E_0 + \Delta_0$ vs. temperature measured with wavelength modulated reflectivity [83T].

Physical property	Numerical value	Experimental conditions	Experimental method, remarks	Ref.
$E_{g,ind}$ $(\Gamma_{15v} - L_{1c})$	2.637(10) eV	78 K	electroabsorption	78K

direct energy gap (in eV):

$E_{g,dir}$ $(\Gamma_{15v} - \Gamma_{1c})$	2.895(4)	0K, extra-polated	photoconductivity	64N
	2.780(2)	300 K		
$E_{gx,dir}$	2.866	0K, extra-polated	excitonic gap, wavelength modulated reflectance	83T

Temperature dependence: Fig. 4; the curve in Fig. 4 can be approximated by $E_0(T) = E_0(0) - 0.1081$ $\cdot (\coth(164/T) - 1)$ (E in eV, T in K).

spin-orbit splitting energy (in eV):
(splitting of Γ_{15v} into Γ_{8v} (upper level) and Γ_{7v} (lower level))

Δ_0	0.080(3)	$100 \cdots 200$ K	wavelength modulated reflectivity	83T

camel's back structure of conduction band edge:
The camel's back structure near the minimum can be described by the formula:

$$E(k) = \hbar^2 k_\parallel^2/2m_l + \hbar^2 k_\perp^2/2m_t - ((\Delta/2)^2 + \Delta_0 \hbar^2 k^2/2m_l)^{1/2}$$

with $k_\parallel$ and $k_\perp$: components of the wave vector parallel and perpendicular to the [100]-direction, respectively, m_t: effective mass perpendicular to the [100]-direction; Δ_0: parameter describing the non-parabolicity; all other parameters are explained in Fig. 2.

Δ	355 meV		fitting of cyclotron resonance data obtained in very high magnetic fields	83M1, 83M2
Δ_0	433 meV			
m_t	$0.25\,m_0$			
m_l	$0.91\,m_0$			
ΔE	3.5 meV		from $\Delta E = (\Delta_0/4)(1 - \Delta/\Delta_0)^2$	
$m_\parallel$	$10.9\,m_0$		from $m_\parallel = m_l(1 - (\Delta/\Delta_0)^2)^{-1}$	
k_m	$0.025\,(2\pi/a)$		from $k_m = (2m_l \Delta E/\hbar^2)^{1/2}$	

effective mass, electrons (in units of m_0):

$m_{n\parallel}^*$	0.254(4)		apparent effective masses from cyclotron resonance data at 119 μm (337 μm) assuming ellipsoidal energy surfaces at X_1 neglecting camel's back structure	83M1
$m_{n\perp}^*$	4.8(5)			

effective masses, holes (in units of m_0):

$m_{p,h}$	0.67(4)	$\parallel [111]$	cyclotron resonance at 1.6 K	72S
$m_{p,l}$	0.17(1)	$\parallel [111]$		
$m_{p,so}$	0.4649		calculated from $k \cdot p$ model	83S1

valence band parameters:

A	−4.20		calculated using $k \cdot p$ theory	75W
B	−1.97			
$|C|$	4.60			

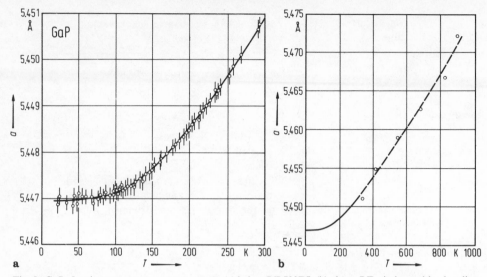

a   **b**

Fig. 5. GaP. Lattice parameter vs. temperature. (a) below RT [83D], (b) above RT, circles and broken line: [72K], solid line: data from (a).

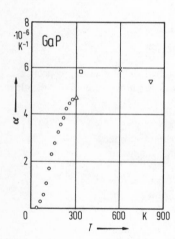

Fig. 6. GaP. Linear thermal expansion coefficient vs. temperature; circles: [83D], other symbols: data taken from literature.

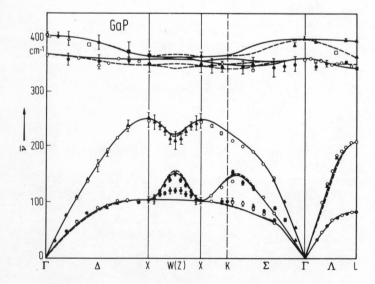

Fig. 7. GaP. Phonon dispersion relations calculated with an eleven parameter rigid ion model. Solid lines calculated to fit neutron diffraction data from [79B] (full symbols). dashed lines calculated to fit data from [68Y] (open symbols [82P]).

Physical property	Numerical value	Experimental conditions	Experimental method, remarks	Ref.

Lattice properties

Structure

GaP I	space group $T_d^2 - F\bar{4}3m$ (zincblende structure)		stable at normal pressure	
GaP II	tetragonal $D_{4h}^{19} - I4_1/amd$ (β-tin structure)		high-pressure phase	82B1

transition pressure:

p_{tr}	215(8) kbar		beginning of phase transition	82B1

lattice parameter:

a	5.4505(2) Å	RT, single crystal		80B

For temperature dependence, see Fig. 5.

linear thermal expansion coefficient: Fig. 6.

density:

d	4.138 g cm^{-3}	300 K		77M

melting point:

T_m	1730(5) K			84T

phonon dispersion relations: Fig. 7.

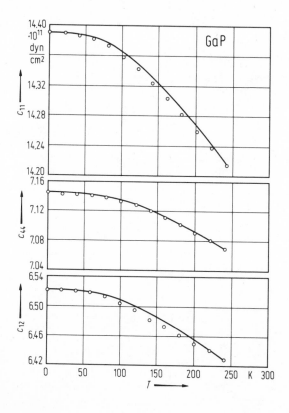

Fig. 8. GaP. Second order elastic moduli vs. temperature [75B].

Physical property	Numerical value	Experimental conditions	Experimental method, remarks	Ref.
phonon frequencies (in THz):				
$\nu(\Gamma)$	12.25(5)	LO	inelastic neutron scattering,	79B
$\nu(X)$	3.13(15)	TA	see also [68Y] and Fig. 7	
	7.5(4)	LA		
	10.65(50)	TO		
	11.0(3)	LO		
$\nu(L)$	2.50(5)	TA		
	10.6(2)	TO		
	12.16(5)	LO		
$\nu(W)$	3.70(10)	A_I		
	4.60(10)	A_{II}		
	6.4(5)	A_{III}		
	10.8(5)	O_I		
second order elastic moduli (in $10^{11}\,\mathrm{dyn\,cm^{-2}}$):				
c_{11}	14.050(28)	300 K	ultrasound ($f = 10/30\,\mathrm{MHz}$),	81Y
c_{12}	6.203(24)		temperature dependence,	
c_{44}	7.033(7)		Fig. 8	
third order elastic moduli (in $10^{12}\,\mathrm{dyn\,cm^{-2}}$):				
c_{111}	−7.37	300 K	ultrasound ($f = 10/30\,\mathrm{MHz}$);	81Y
c_{112}	−4.74		hydrostatic pressure up to	
c_{123}	−1.31		1.5 kbar, uniaxial pressure	
c_{144}	−1.07		up to 30 bar	
c_{155}	−2.34			
c_{456}	−0.62			

Transport properties

Intrinsic conductivity in GaP occurs even in high purity samples only above 500 °C. Thus, the transport properties are in general determined by the properties of impurities and lattice defects.

As in GaAs (see section 2.10) *semi-insulating GaP* can be obtained by doping with shallow donors and acceptors as well as deep centers. Resistivities at RT are of the order of $10^8 \cdots 10^{11}\,\Omega\,\mathrm{cm}$.

carrier mobilities:				
$\mu_{H,n}$	160 cm²/Vs	RT, LPE grown layers	maximum mobility, temperature dependence $\propto T^{-1.7}$	83K2
$\mu_{H,p}$	135 cm²/Vs		maximum mobility, temperature dependence $\propto T^{-2.3}$	

Temperature dependence (LPE grown layers): Fig. 9.

thermal conductivity: Fig. 10.

Optical properties

dielectric constants:

$\varepsilon(0)$	11.11	300 K	$\varepsilon(0)$: low-frequency capacitance	83S2
	10.86	75.7 K	measurements	
$\varepsilon(\infty)$	9.11	300 K	$\varepsilon(\infty)$: derived from refractive index data	

refractive index (see also below): range $1.5 \cdots 2.7\,\mathrm{eV}$: Fig. 11.

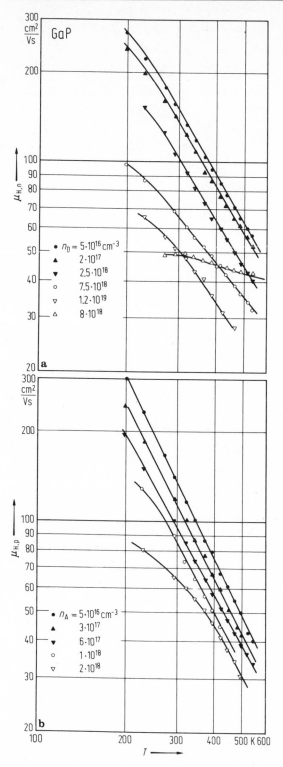

Fig. 9. GaP. Hall mobility in n-type (a) and p-type (b) LPE-grown layers vs. temperature [83K2].

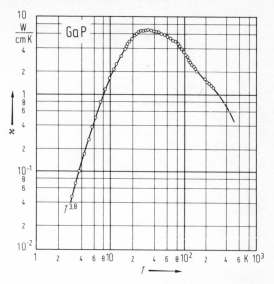

Fig. 10. GaP. Thermal conductivity vs. temperature for a p-type sample [69M3].

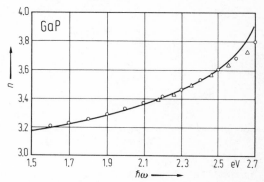

Fig. 11. GaP. Refractive index vs. photon energy in the range 1.5···2.7 eV. Circles and solid curve: calculated, triangles from [82B3][85C].

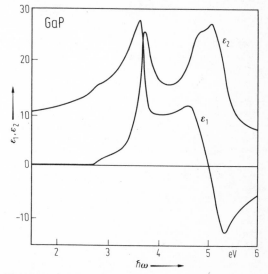

Fig. 12. GaP. Real and imaginary parts of the dielectric constant vs. photon energy [83A].

optical constants

real and imaginary parts of the dielectric constant measured by spectroscopical ellipsometry; n, k, R, K calculated from these data [83A]; see also Fig. 12.

$\hbar\omega$ [eV]	ε_1	ε_2	n	k	R	K [10^3 cm^{-1}]
1.5	10.102	0.000	3.178	0.000	0.272	0.00
2.0	11.114	0.000	3.334	0.000	0.290	0.00
2.5	12.996	0.046	3.605	0.006	0.320	1.63
3.0	16.601	1.832	4.081	0.224	0.369	68.26
3.5	24.833	8.268	5.050	0.819	0.458	290.40
4.0	9.652	16.454	3.790	2.171	0.452	880.10
4.5	11.073	17.343	3.978	2.180	0.461	994.27
5.0	0.218	26.580	3.661	3.631	0.580	1839.99
5.5	−10.266	10.974	1.543	3.556	0.677	1982.53
6.0	−5.521	7.041	1.309	2.690	0.583	1635.71

Impurities and defects

diffusion coefficients of impurities in GaP

Impurity	D_0 [cm^2 s^{-1}]	Q [eV]	Remarks	Ref.
Au	8	2.5 ± 0.3	from A face, $c^{eq} = 10^{16}$ cm^{-3} (1250 °C)	78D
Au	20	2.4 ± 0.2	from B face, $c^{eq} = 10^{16}$ cm^{-3} (1300 °C)	78D
Be			concentration dependent profiles	72I
S	$3.2 \cdot 10^3$	4.7	radiotracer	70Y
Zn			chemical diffusion profiles complex,	64C

shallow impurities

The effect of deviations from the ellipsoidal form of the conduction band (camel's back structure) on the donor states becomes serious since $E_{min}(\Delta) - E(X)$ is comparable to the binding energies of the donor excited states. Each state splits into two subcomponents involving a symmetric and antisymmetric contribution from the individual sub-minima [77K]. There is still some uncertainty over the exact classification of certain excited states of P-site donors and their magnetic properties which deviate significantly from the simple form of ellipsoidal minima [77C1].

Binding energies of donors

Impurity	E_b [meV]	T [K]	Remarks	Ref.
Theory	59(2)		effective mass theory	77K
O_P	897(1)	2	infrared photoluminescence (consistent with threshold of photo-neutralization spectra [78B, 78M1])	68D
S_P	107(1)	20	infrared absorption	78K
Se_P	105(1)	2	donor-acceptor pair spectra	73D2
Te_P	92.6(1.0)	20	infrared absorption	78K
Si_{Ga}	85(1)	20	infrared absorption	78K
Ge_{Ga}	204(1)	2	donor-acceptor spectra	73D3
Sn_{Ga}	72(1)		donor-acceptor spectra	73D3
Li_P	91(1)	2	bound exciton 'two-electron' replicas	73D1
Li_{Ga}	61(2)	2	unresolved donor acceptor pair spectra	

binding energies of acceptors

Impurity	E_b [meV]	T [K]	Remarks	Ref.
Theory	45.3 or 56.3		effective mass theory according to [74B], re-evaluated for $\varepsilon = 11.02$	78K
C_P	54.3(5)	2	donor-acceptor pair spectra (all data from [73D3] revised upward by 77.9 meV because of the increase in E_q discussed in [78K] and [78B], see also [78S] where an increase of 8.3 meV is preferred)	73D2
Si_P	210(1)			
Ge_P	265(1)			
Cu_P	530(30)	4		73G
	580	77	photoconductivity, absorption (the complex and ill-understood behavior of GaP:Cu in thermally stimulated current measurements has been noted by [78R])	71O,
	650			74K
Be_{Ga}	56.6(5)	2	photoconductivity, absorption	73D2
Mg_{Ga}	59.9(5)			
Zn_{Ga}	69.7(5)	20	infrared absorption	78K
Cd_{Ga}	102.2(5)			
X	52(1)	2	donor-acceptor pair spectra	73D2

transition metal impurities

Energy levels related to isolated, substitutional transition metal impurities. ("$+$" above valence band edge, "$-$" below conduction band edge).

Impurity	E[eV]	Type	Remarks	T[K]	Ref.
Ti	$-0.50(2)$	a	DLTS and DLOS	$77 \cdots 350$	87R
	$+1.0(2)$	d			
V	-0.58	a	Temperature dependent Hall effect		86U
	$+0.20(5)$	d	Electrical + optical measurements		89U
Cr	-1.2	1st a	Reinterpretation of		81C
	-0.5	2nd a	photo-ESR data [78K1]		81C
Mn	$+(0.4 \cdots 0.43)$	a	Temperature dependent Hall effect		76E
Fe	$+0.82$	1st a	DLTS		85B
	-0.26	2nd a	DLTS		85B
Co	$+0.41$	1st a	Temperature dependent Hall and resistivity measurements		66L
	-0.33	2nd a			83J
Ni	$+0.51$	1st a	DLTS, σ_p^0 photocapacitance and photo-ESR		84P
	-0.82	2nd a	DLTS and transient capacitance		84Y

References for 2.9

64C Chang, L.L., Pearson, G.L.: J. Appl. Phys. 35 (1964) 374.
64N Nelson, D.F., Johnson, L.F., Gershenzon, M.: Phys. Rev. 135 (1964) A1399.
66L Loescher, D.H., Allen, J.W., Pearson, G.L.: J. Phys. Soc. Jpn. 21 Suppl. (1966) 239.
68D Dean, P.J., Henry, C.H.: Phys. Rev. 176 (1968) 928.

68Y   Yarnell, J.L., Warren, J.L., Wenzel, R.G., Dean, P.J.: in "Neutron Inelastic Scattering Symposion", Copenhagen 1968, vol. 1, IAEA, Vienna 1968, p. 301.
69M   Muzhdaba, V.M., Nashel'skii, A.Ya., Tamarin, P.V., Shalyt, S.S.: Sov. Phys. Solid State (English Transl.) 10 (1969) 2265; Fiz. Tverd. Tela 10 (1968) 2866.
69P   Panish, M.B., Casey, H.C.: J. Appl. Phys. 40 (1969) 163.
70Y   Young, A.B.Y., Pearson, G.L.: J. Phys. Chem. Solids 31 (1970) 517.
71O   Olsson, R.: Phys. Status Solidi (b) 46 (1971) 163.
72I   Ilegems, M., O'Mara, W.C.: J. Appl. Phys. 43 (1972) 1190.
72K   Kudman, I., Pfaff, R.J.: J. Appl. Phys. 43 (1972) 3760.
72S   Schwerdtfeger, C.P.: Solid State Commun. 11 (1972) 779.
73D1 Dean, P.J.: Luminescence of Crystals, Molecules and Solutions, Williams, F.E. (ed.), New York: Plenum Press 1973, p. 538.
73D2 Dean, P.J.: Progress in Solid State Chemistry, Vol. 8, McCaldin, J.O., Somorjai, G. (eds.), New York: Pergamon Press 1973, p. 1.
73G   Grimmeiss, H.G., Monemar, B.: Phys. Status Solidi (a) 19 (1973) 505.
74K   Kopylov, A.A., Pikhtin, A.N.: Sov. Phys. Solid State (English Transl.) 16 (1974) 1837; Fiz. Tverd. Tela 16 (1974) 1837.
74L   Ley, L., Pollak, R.A., McFeely, R.R., Kowalczyk, S.P.: Phys. Rev. B9 (1974) 600.
75B   Boyle, W.F., Sladek, R.J.: Phys. Rev. B11 (1975) 2933.
75W   Wiley, J.D.: in "Semiconductors and Semimetals", Vol. 10, R.K. Willardson and A.C. Beer eds., Academic Press, New York and London 1975.
76C   Chelikowsky, J.R., Cohen, M.L.: Phys. Rev. B14 (1976) 556.
76E   Evwaraye, A.O., Woodbury, H.H.: J. Appl. Phys. 47 (1976) 1595.
76S   Street, R.A., Senske, W.: Phys. Rev. Lett. 37 (1976) 1292.
77K   Kopylov, A.A., Pikhtin A.N.: Sov. Phys. Semicond. (English Transl.) 11 (1977) 510; Fiz. Tekh. Poluprovodn. 11 (1977) 867.
78D   Dzhafarov, T.D., Litvin, A.A., Khudyakov, S.V.: Sov. Phys. Solid State 20 (1978) 152.
77M   Merrill, L.: J. Phys. Chem. Ref. Data 6 (1977) 1205.
78H   Humphreys, R.G., Rössler, U., Cardona, M.: Phys. Rev. B18 (1978) 5590.
78K   Kopylov, A.A., Pikhtin, A.N.: Solid State Commun. 26 (1978) 735.
79B   Borcherds, P.H., Kung, K., Alfreys, G.F., Hall, R.L.: J. Phys. C12 (1979) 4699.
80B   Bessolov, V.N., Dedegkaev, T.T., Efimov, A.N., Kartenko, N.F., Yakovlev, Yu.P.: Sov. Phys. Solid State (English Transl.) 22 (1980) 1652; Fiz. Tverd. Tela 22 (1980) 2834.
81C   Clerjaud, B., Gendron, F., Porte, C.: Appl. Phys. Lett. 38 (1981) 212.
81Y   Yogurtcu, Y.K., Miller, A.J., Saunders, G.A.: J. Phys. Chem. Solids 42 (1981) 49.
82B1 Baublitz, M., Ruoff, A.L.: J. Appl. Phys. 53 (1982) 6179.
82B2 Burkhard, H., Dinges, H.W., Kuphal, E.: J. Appl. Phys. 53 (1982) 655.
82P   Patel, C., Sherman, W.F., Wilkinson, G.R.: Phys. Status Solidi (b) 111 (1982) 649.
83A   Aspnes, D.E., Studna, A.A: Phys. Rev. B27 (1983) 985.
83D   Deus, P., Voland, U., Schneider, H.A.: Phys. Status Solidi (a) 80 (1983) K29.
83J   Jezewski, M., Baranowski, J.M.: 4th "Lund" Int. Conf. on Deep Level Impurities in Semicond., Eger Hungary, 1983, unpublished.
83K   Kao, Y.C., Eknoyan, O.: J. Appl. Phys. 54 (1983) 2468.
83M1 Miura, N., Kido, G., Suekane, M., Chikazumi, S.: J. Phys. Soc. Jpn. 52 (1983) 2838.
83M2 Miura, N., Kido, G., Suekane, M., Chikazumi, S.: Physica 117B&118B (1983) 66.
83S1 Sharma, A.C., Ravindra, N.M., Auluck, S., Srivastava, V.K.: Phys. Status Solidi (b) 120 (1983) 715.
83S2 Samara, G.A.: Phys. Rev. B27 (1983) 3494.
83T   Takizawa, T.: J. Phys. Soc. Jpn. 52 (1983) 1057.
84P   Peaker, A.R., Kaufmann, U., Zhan-Guo Wang, Wörner, R., Hamilton, B., Grimmeiss, H.G.: J. Phys. C17 (1984) 6161.
84S   Solal, F., Jezequel, G., Houzay, F., Barski, A., Pinchaux, R.: Solid State commun. 52 (1984) 37 and private communication from G. Jezequel.
84T   Tmar, M., Gabriel, A., Chatillon, C., Ansara, I.: J. Crystal Growth 68 (1984) 557.
84Y   Yang, X.Z., Samuelson, L., Grimmeiss, H.G.: J. Phys. C17 (1984) 6521.
85B   Brehme, S.: J. Phys. C18 (1985) L319.
85C   Campi, D., Papuzza, C.: J. Appl. Phys. 57 (1985) 1305.
86U   Ulrici, W., Eaves, L., Friedland, K., Halliday, D.P., Kreissl, J.: Defects in Semiconductors, Proc. 14th Internat. Conf. Defects in Semicond., Paris (1986), von Bardeleben, H.J. (ed.), Materials Science Forum, Vol. 10···12, Trans. Tech. Publications, Switzerland, 1986, p. 639.
87R   Roura, P., Bremond, G., Nouailhat, A., Guillot, G., Ulrici, W.: Appl. Phys. Lett. 51 (1987) 1696.
89U   Ulrici, W., Kreissl, J., Hayes, D.G., Eaves, L., Friedland, K.: Materials Science Forum, Vol. 38···41, Trans. Tech. Publications, Switzerland, 1989, p 875.

Physical property	Numerical value	Experimental conditions	Experimental method, remarks	Ref.

2.10 Gallium arsenide (GaAs)

Electronic properties

band structure: Fig. 1 (Brillouin zone: section 1.1, Fig. 2).

GaAs is a direct gap semiconductor. The *conduction band* minimum is situated at Γ (symmetry Γ_8). Higher sets of minima at L and near X are important for optical as well as transport properties. The X-minimum has most probably a camel's back structure. The *valence bands* show the topology characteristic for all zincblende semiconductors.

energies of symmetry points of the band structure (relative to the top of the valence band) (in eV):

$E(\Gamma_{6v})$	-12.55	$E(\Gamma_{1v})$	-13.1	first row: designation of
$E(\Gamma_{7v})$	$-0.35\}$			symmetry point in double
$E(\Gamma_{8v})$	$0.00\}$	$E(\Gamma_{15v})$	0.00	group notation, data
$E(\Gamma_{6c})$	1.51	$E(\Gamma_{1c})$	1.632	according to a nonlocal
$E(\Gamma_{7c})$	$4.55\}$			pseudopotential calculation
$E(\Gamma_{8c})$	$4.71\}$	$E(\Gamma_{15c})$	4.716	[76C] (see Fig. 1)
$E(\Gamma_{6c})$	—	$E(\Gamma_{1c})$	8.33	second row: designation of
$E(X_{6v})$	-9.83	$E(X_{1v})$	-10.75	symmetry point in single
$E(X_{6v})$	-6.88	$E(X_{3v})$	-6.70	group notation, data
$E(X_{6v})$	$-2.99\}$			from angle resolved photo-
$E(X_{7v})$	$-2.89\}$	$E(X_{5v})$	-2.80	emission [80C] and photo-
$E(X_{6c})$	2.03	$E(X_{1c})$	2.18	electron spectroscopy [85S]
$E(X_{7c})$	2.38	$E(X_{3c})$	2.58	
$E(L_{6v})$	-10.60	$E(L_{1v})$	-11.24	
$E(L_{6v})$	-6.83	$E(L_{1v})$	6.70	
$E(L_{6v})$	$-1.42\}$			
$E(L_{4,5v})$	$-1.20\}$	$E(L_{3v})$	-1.30	
$E(L_{6c})$	1.82	$E(L_{1c})$	1.85	
$E(L_{6c})$	$5.47\}$			
$E(L_{45c})$	$5.52\}$	$E(L_{3c})$	—	

direct gap (in eV):

$E_{g,dir}$	1.51914	0 K (extrapol.)	fitting of photoluminescence	84S
$(\Gamma_{8v} - \Gamma_{6c})$			data	
	$1.424(1)$	300 K	differentiated reflectivity	74S1
$E_{g,th}$	1.604	0 K, extrapol.	intrinsic carrier concentration	82B3

temperature dependence of $E_{g,dir}$:
Empirical relation: $E_{g,dir}(T) = 1.519 - 5.408 \cdot 10^{-4} T^2/(T+204)$ eV (T in K) according to [82B1]. See also Fig. 2.

exciton energies (in eV):

$E(1S, \Gamma_{5T})$	1.5150	1.8 K	reflectance	83S1
$E(1S, \Gamma_{5L})$	1.51544	10 K	resonant Brillouin scattering	83S2

higher conduction band minima (energy above Γ_{6c} minimum) (in eV):

$\Delta E_{\Gamma L}(\Gamma_{6c} - L_{6c})$	$0.300(20)$	120 K	photoemission	85D
$\Delta E_{\Gamma X}(\Gamma_{6c} - X_{6c})$	$0.460(20)$			

spin-orbit splitting energy:

$\Delta_0(\Gamma_{7v} - \Gamma_{8v})$	$0.341(1)$ eV	4.2 K	electroreflectance	73A

Physical property	Numerical value	Experimental conditions	Experimental method, remarks	Ref.
critical point energies (in eV):				
$E_0 + \Delta_0$	1.859(1)	4.2 K	electroreflectance; assigned transition: $\Gamma_{7v} - \Gamma_{6c}$	73A
E_0'	4.488(10)		$\Gamma_{8v} - \Gamma_{7c}$	
E_0'	4.529(10)		$\Delta_{5v} - \Delta_{5c}$	
$E_0' + \Delta_0'$	4.659(10)		$\Gamma_{8v} - \Gamma_{8c}$	
$E_0' + \Delta_0''$	4.712(10)		$\Delta_{5v} - \Delta_{5c}$	
$E_0' + \Delta_0' + \Delta_0$	5.014(15)		$\Gamma_{7v} - \Gamma_{8c}$	
E_1	3.043(9)		$\Lambda_{4,5v} - \Lambda_{6c}$	
$E_1 + \Delta_1$	3.2636(1)		$\Lambda_{6v} - \Lambda_{6c}$	
E_2	5.137(10)		at Σ	
	4.937(10)		$X_{7v} - X_{6c}$	
	5.014(10)		$X_{6v} - X_{6c}$	
	5.339(10)		$X_{7v} - X_{7c}$	
	5.415(15)		$X_{6v} - X_{7c}$	

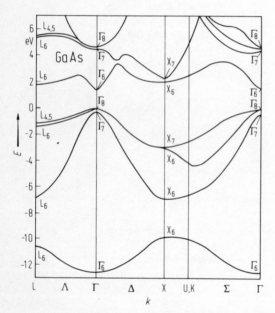

Fig.1. GaAs. Band structure obtained by a non-local pseudopotential calculation [76C].

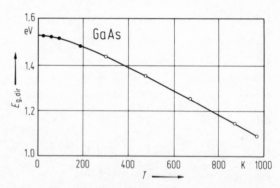

Fig. 2. GaAs. Direct energy gap vs. temperature. Data from two authors [69P].

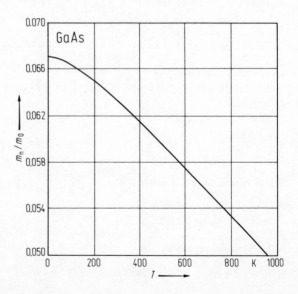

Fig. 3. GaAs. Effective electron mass vs. temperature obtained by a $k \cdot p$ calculation using experimental values for the input parameters [82B1].

Physical property	Numerical value	Experimental conditions	Experimental method, remarks	Ref.
effective mass, electrons (in units of m_0):				
$m_n(\Gamma)$	0.067	0 K	recommended band edge density of states mass	82B1
	0.063	300 K		
	0.0650(5)		deduced from cyclotron resonance corrected for polaron effects	83L1

For temperature dependence of m_n, see Fig. 3.

camel's back structure of conduction band edge:				
ΔE	9.3 meV		for definition of the parameters, see Fig. 2 of section 2.9; values extrapolated from GaP using a $k \cdot p$ model	85K
Δ	304 meV			
k_m	0.102 $(2\pi/a)$			
m_1	1.8 m_0			
m_t	0.257 m_0			
$m_{X,ds}$	0.85 m_0		density of states mass at X neglecting camel's back structure	82B1

g-factor of electrons:				
g_c	$-0.44(5)$	4 K	piezospectroscopy	76S2

effective masses, holes (in units of m_0):				
$m_{p,h}$	0.51(2)	$T < 100$ K	recommended density of states mass values from discussion of various literature data	82B1
	0.50	300 K		
$m_{p,1}$	0.082(4)	$T < 100$ K		
	0.076	300 K		
m_{so}	0.154	$T < 100$ K		
	0.145	300 K		
$m_{p,ds}$	0.53		combined for heavy and light hole bands	

valence band parameters:				
A	$-6.98(45)$	77 K	cyclotron resonance	76S2
B	$-4.5(2)$			
$\lvert C \rvert$	6.2(10)			

Lattice properties

Structure

GaAs I	space group T_d^2–F$\bar{4}$3m (zincblende structure)	stable at normal pressure	
GaAs II	orthorhombic distortion of NaCl structure space group probably D_{2h}^1–Pmmm	high-pressure phase	82B2

transition pressure:

p_{tr}	172(7) kbar		beginning of phase transition	82B2

Physical property	Numerical value	Experimental conditions	Experimental method, remarks	Ref.
lattice parameter:				
a	5.65325(2) Å	300 K	X-ray, single crystal	75M2
For temperature dependence, see Fig. 4.				
linear thermal expansion coefficient: Fig. 5.				
density:				
d	5.3176(3) g cm^{-3}	298.15 K	Archimedean	65S2

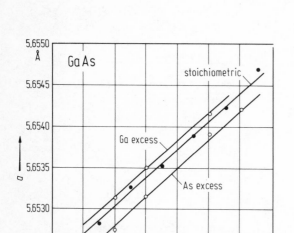

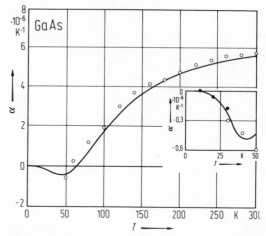

Fig. 5. GaAs. Coefficient of linear thermal expansion vs. temperature. Theoretical curve compared with experimental data [82S1].

Fig. 4. GaAs. Lattice parameter vs. temperature for stoichiometric crystals and crystals with Ga and As excess, respectively [65S2].

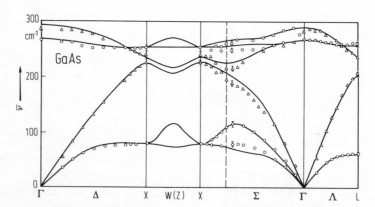

Fig. 6. GaAs. Phonon dispersion relations calculated with an eleven-parameter rigid-ion model. Experimental points from neutron diffraction [84P2].

Physical property	Numerical value	Experimental conditions	Experimental method, remarks	Ref.

melting point:

T_m 1513 K

phonon dispersion relations: Fig. 6.

phonon wavenumbers (in cm^{-1}):

$\bar{\nu}_{LO}(\Gamma)$	285.0(67)		neutron scattering	84P
$\bar{\nu}_{TO}(\Gamma)$	267.3(27)			
$\bar{\nu}_{LO}(X)$	240.7(50)			
$\bar{\nu}_{LA}(X)$	226.7(20)			
$\bar{\nu}_{TO}(X)$	252.0(30)			
$\bar{\nu}_{TA}(X)$	78.7(5)			
$\bar{\nu}_{LO}(L)$	238.3(23)			
$\bar{\nu}_{LA}(L)$	208.7(33)			
$\bar{\nu}_{TO}(L)$	261.3(40)			
$\bar{\nu}_{TA}(L)$	62.0(7)			

second order elastic moduli (in $10^{11}\,dyn\,cm^{-2}$):

c_{11}	11.9(1)	300 K	recommended values from a	82B1
	12.21(3)	77 K	discussion of several published	
	11.26	0 K	data	
c_{12}	5.38(1)	300 K	$T = 0\,K$ data: extrapolated	
	5.66(3)	77 K		
	5.71	0 K		
	5.95(1)	300 K		
c_{44}	5.99(1)	77 K		
	6.00	0 K		

third order elastic moduli (in $10^{12}\,dyn\,cm^{-2}$):

c_{111}	−6.75(20)	RT,	ultrasound	66D
c_{112}	−4.02(10)	$n < 10^{17}\,cm^{-3}$		
c_{123}	−0.04(10)			
c_{144}	−0.70(10)			
c_{166}	−3.20(20)			
c_{456}	−0.69(3)			

Transport properties

At low fields, the electrons occupy the Γ_6 minima at the zone center. The dominating scattering process at RT is polar optical scattering, while below 60 K the most important contribution to the lattice mobility is made by piezoelectric scattering. At RT, the hole mobility of samples with $p < 5 \cdot 10^{15}\,cm^{-3}$ is governed by lattice scattering alone.

intrinsic carrier concentration (in cm^{-3}):

n_i	$2.1 \cdot 10^6$	300 K		82B3

The temperature dependence can be expressed by

$$n_i(T) = 1.05 \cdot 10^{16}\, T^{3/2} \exp(-1.604/2kT)$$

(T in K; kT in eV) for the range $33\,K < T < 475\,K$. For the range $250\,K < T < 1500\,K$ and for comparison of various literature data, see Fig. 7.

Physical property	Numerical value	Experimental conditions	Experimental method, remarks	Ref.

At RT a fraction of about $4 \cdot 10^{-4}$ of the conduction electrons fills the L_6-bands; the fraction of the conduction electrons in the X_6-bands is negligible ($< 10^{-6}$). At 400 K the L_6-bands begin to hold an appreciable fraction of carriers and at 800 K the X_6-bands also take over parts of the electron population. At 100 K only 30% of all conduction electrons reside in the Γ_6-band. Thus, above RT the "one-band" approximation $n = 1/|R_H e|$ is of limited accuracy [80N].

electrical conductivity:

Since n_i (300 K) $= 2.1 \cdot 10^6$ cm^{-3} the electrical conductivity in GaAs is in general determined by charge carriers provided by impurities. Only by certain doping conditions (e.g. compensation of shallow acceptor levels by donor impurities and overcompensation by deep Cr levels) the Fermi level can be pinned in the middle of the energy gap and nearly intrinsic conduction can be achieved (*semi-insulating GaAs*, $\varrho = 10^5 \cdots 10^9 \, \Omega$ cm).

typical data for semi-insulating GaAs:

Physical property	Numerical value	Experimental conditions	Experimental method, remarks	Ref.
σ	$2.38 \cdot 10^{-9} \, \Omega^{-1}$ cm^{-1}	RT, undoped sample	n and μ_n obtained from R_H and σ by a mixed conduction analysis	83W
n	$3.73 \cdot 10^6$ cm^{-3}			
$\mu_{H,n}$	3970 cm^2/Vs			

electron mobility (in cm^2/Vs):

Physical property	Numerical value	Experimental conditions	Experimental method, remarks	Ref.
$\mu_{H,n}$	9200	300 K	epitaxial layer	82L1
	$2 \cdot 10^5$	77 K	$n(77 \text{ K}) = n(300 \text{ K})$	
	$3.76 \cdot 10^5$	41 K	$= 1.3 \cdot 10^{13}$ cm^{-3}	
	$9400 (300/T)^{2.3}$		(T in K) in weak fields	83B1
$\mu_{dr,n}$	$8000 (300/T)^{2.3}$		fairly close to RT	

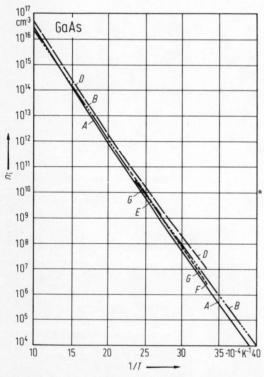

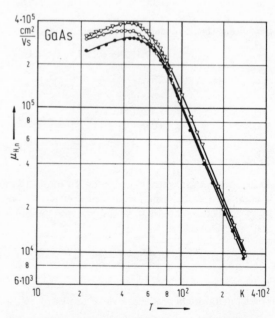

Fig. 7. GaAs. Intrinsic carrier concentration vs. reciprocal temperature for the range $250 \cdots 1000$ K. The solid curve (A) is the result of a critical discussion of the literature data.- Curves (B) $\cdots$ (G) represent data from other sources [82B].

Fig. 8. GaAs. Electron Hall mobility vs. temperature for three high purity epitaxial layers ($n(300 \text{ K}) \simeq 3 \cdot 10^{13}$ cm^{-3}). Solid curves: theoretical [82L1].

Physical property	Numerical value	Experimental conditions	Experimental method, remarks	Ref.

For temperature dependence in the range 20 K ⋯ 300 K, see Fig. 8. Separation of the mobility into contributions of several bands: Fig. 9.

hole mobility (in cm^2/Vs):

$\mu_{H,p}$	$400(300/T)^{2.3}$	near 300 K	(T in K)	75W
	402	300 K	theoretical, see Fig. 10	83L2
$\mu_{dr,p}$	$320(300/T)^{2.3}$			75W
heavy holes:	130	300 K	theoretical, see Fig. 10	83L2
light holes:	1060			

thermal conductivity: Fig. 11.

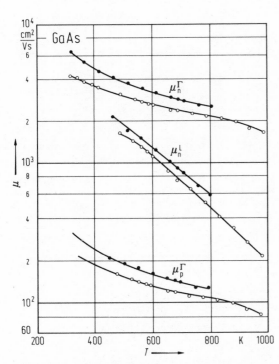

Fig. 9. GaAs. Carrier mobilities in the Γ- and L-minima of the conduction bands and in the Γ$_8$ valence bands for two samples [80N].

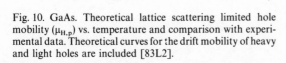

Fig. 10. GaAs. Theoretical lattice scattering limited hole mobility ($\mu_{H,p}$) vs. temperature and comparison with experimental data. Theoretical curves for the drift mobility of heavy and light holes are included [83L2].

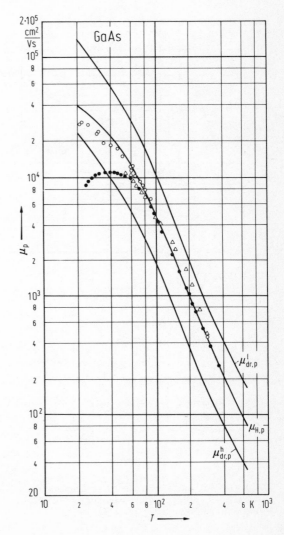

Physical property	Numerical value	Experimental conditions	Experimental method, remarks	Ref.

Optical properties

dielectric constants:

$\varepsilon(0)$	$12.40\cdot(1 + 1.2\cdot10^{-4}T)$ $(T \text{ in K})$		recommended value and linearized temperature dependence below 600 K	82B1
$\varepsilon(\infty)$	$10.60\cdot(1 + 9.0\cdot10^{-5}T)$			

refractive index (see also below):

n	$3.255\cdot(1 + 4.5\cdot10^{-5}T)$	$(T \text{ in K})$		82B1

Dependence on photon energy: range $1\cdots1.7$ eV: Fig. 12.

In the range $0.3\cdots1.4$ eV various published data can be fitted by

$$n^2 = 7.10 + 3.78/(1 - 0.18(\hbar\omega)^2) \quad (\hbar\omega \text{ in eV}) \qquad\qquad 82B1$$

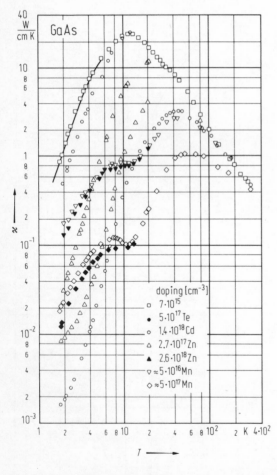

Fig. 11. GaAs. Thermal conductivity vs. temperature for seven samples with impurity concentrations between $7\cdot10^{15}$ cm^{-3} and $2.6\cdot10^{18}$ cm^{-3} [64H2].

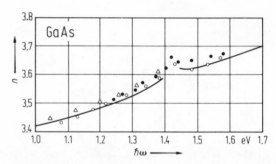

Fig. 12. GaAs. Refractive index vs. photon energy in the range $1.0\cdots1.7$ eV [85C], solid line: calculated, symbols: experimental data from three sources.

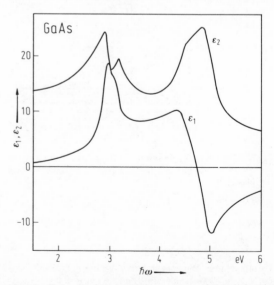

Fig. 13. GaAs. Real and imaginary parts of the dielectric constant vs. photon energy [83A].

optical constants:

real and imaginary parts of the dielectric constant measured by spectroscopical ellipsometry; n, k, R, K calculated from these data [83A]; see also Fig. 13.

$\hbar\omega$ [eV]	ε_1	ε_2	n	k	R	K [10^3 cm^{-1}]
1.5	13.435	0.589	3.666	0.080	0.327	12.21
2.0	14.991	1.637	3.878	0.211	0.349	42.79
2.5	18.579	3.821	4.333	0.441	0.395	111.74
3.0	16.536	17.571	4.509	1.948	0.472	592.48
3.5	8.413	14.216	3.531	2.013	0.425	714.20
4.0	9.279	13.832	3.601	1.920	0.421	778.65
4.5	6.797	22.845	3.913	2.919	0.521	1331.28
5.0	-11.515	18.563	2.273	4.084	0.668	2069.81
5.5	-6.705	8.123	1.383	2.936	0.613	1636.68
6.0	-4.511	6.250	1.264	2.472	0.550	1503.20

Impurities and defects

solubility of impurities in GaAs

Impurity	c^{eq} [cm^{-3}]	T [°C]	Remarks	Ref.
Ag	$5\cdot10^{20}$	1000	radiotracer	64B
Au	$1.5\cdot10^{17}$	835	radiatracer	68K
C	$1.9\cdot10^{19}$	1100	^{14}C, mass spectrometry	78B
Cd	$\approx 2\cdot10^{18}$	1000	sheet resistance	64K2
Cr	$\approx 5\cdot10^{16}$	1100	radiotracer, possible As and interstitial sites	79T
Cu	$7\cdot10^{18}$ (max)	1100	radiotracer	64H
	$1.5\cdot10^{16}$	<700		
Ge	$>2\cdot10^{18}$	melt growth	lattice parameter, electrical activity	75M1
Hg	$5\cdot10^{17}$	1000		68K
Mn	$10^{18}\cdots10^{19}$	825	radiotracer	66L
O	$4\cdot10^{19}$ (max)	1100	mass spectrometer retrograde solubility	77B
Pb	$4\cdot10^{18}$	melt growth	lattice parameter, electrical activity	75M1
S	$1.6\cdot10^{18}$	900		60R
	$4\cdot10^{18}$	melt growth	lattice parameter, electrical activity	75M1
Se	$c_{Se}^{eq} = 9.5\cdot10^{23}\exp(-1.23(2)\,\mathrm{eV}/kT)\,\mathrm{cm}^{-3}$			78L
	$2\cdot10^{18}$	melt growth		75M1
Si	$>4.7\cdot10^{18}$	melt growth	n-type (Si$_{Ga}$?)	80F
	$>3\cdot10^{18}$	melt growth	p-type (Si$_{As}$?)	80F
	$6\cdot10^{18}$	LPE growth $856\cdots750$	p-type (both Si$_{As}$ and Si$_{Ga}$?)	80F
Sn	$>1\cdot10^{19}$	LPE growth $856\cdots750$		75M1
Tc	$1\cdot10^{19}$	melt growth	lattice parameter, electrical activity	75M1
	$4.5\cdot10^{18}$	880	electrical activity, lattice parameter after anneal	79D
Zn	$4\cdot10^{20}$	1000	radiotracer	64C2
	$3.8\cdot10^{20}$	1000	radiotracer	76T

diffusion coefficients

Element	$D_0 [\text{cm}^2\,\text{s}^{-1}]$	$Q [\text{eV}]$	Remarks	Ref.
Self-diffusion coefficients				
Ga	$4 \cdot 10^{-5}$	2.6	radiotracer	83P
As	$5.5 \cdot 10^{-4}$	3.0	radiotracer, arsenic pressure dependence	83P
Impurity diffusion coefficients				
Ag	$4.0 \cdot 10^{-4}$	0.80	radiotracer	64B
	$2.5 \cdot 10^{1}$	2.27	radiotracer	73C
	complex profiles		radiotracer ($\sqrt{Dt} \simeq$ slice thickness)	81T
Au	$1.0 \cdot 10^{-3}$	1.0	radiotracer	64S
	$2.9 \cdot 10^{1}$	2.64	radiotracer	73C
Be	$7.3 \cdot 10^{-6}$	1.2	incremental sheet resistance	66P
Cd	$5.0 \cdot 10^{-2}$	2.43	radiotracer	60G
	concentration dependent		radiotracer	69S
Cr	$4.3 \cdot 10^{3}$	3.4	radiotracer	73C
	complex profiles dependent on As pressure		radiotracer ($\sqrt{Dt} \simeq$ slice thickness)	79T
	$6.3 \cdot 10^{5}$	3.4	outdiffusion	80K1
	$D(800\,°C) = 6.7 \cdot 10^{-12}$		SIMS	80W
	$8.53 \cdot 10^{4}$	3.53	outdiffusion, SIMS	82M
Cu	$3.0 \cdot 10^{-2}$	0.53	radiotracer	64H
Hg	$D(1000\,°C) = 5 \cdot 10^{-14}$			73C
In	$D(1000\,°C) = 7 \cdot 10^{-11}$			64K
Mg	$2.6 \cdot 10^{-2}$	2.7	pn-junction, SIMS	65M
				82S2
Mn	$6.5 \cdot 10^{-1}$	2.49	radiotracer, dependent on As pressure	65S1
O	$2.0 \cdot 10^{-3}$	1.1	outdiffusion	69R
S	$1.2 \cdot 10^{-4}$	1.8	pn-junction	61V
	dependence on As pressure		radiotracer	81T
Se	$3.0 \cdot 10^{3}$	4.16	radiotracer, Ga_2Se_3 layer formed	61G
Si	0.11	2.5	SIMS and differential Hall effect	84G
			SIMS, defect model	87K
Sn	$3.5 \cdot 10^{-2}$	2.7	radiotracer	63F
	3.2	3.3	radiotracer, undoped material, excess As pressure	78T
	$9.43 \cdot 10^{-8}$	1.9	radiotracer, n-type material $n \approx 2 \cdot 10^{18}\,\text{cm}^{-3}$	78T
Te	$D(1000\,°C) = 10^{-13}$			73C
	$D(1100\,°C) = 2 \cdot 10^{-12}$			
Tm	$2.3 \cdot 10^{-16}$	$(-)1.0$		64C1

Zinc diffusion data in GaAs

By far the most attention has been paid to the diffusion of zinc in gallium arsenide, because of its importance as a p-type dopant. The complex nature of the dependence of zinc diffusivity on the zinc concentration, the arsenic pressure, and the temperature has shown that it is absolutely vital to consider the ternary nature of the system in the interpretation and characterization of this dopant, and we cannot consider the compound as one element in a binary system.

shallow impurities

Impurity	E_b[meV]	T[K]	Remarks	Ref.

shallow donors, binding energies (in meV)

The binding energies in the following table are calculated from the energetic positions of the 1S–2P transitions as observed in photoconductivity plus the binding energy of the 2P-state of 1.429 meV as calculated in [71S4]. The energy positions of the 1S–2P transitions without magnetic field are extrapolated from magnetic field experiments

Theory	5.715		donor effective mass calculation	71S2
Pb, 'A' = ?	5.752	4.2	far infrared photoconductivity	76S1,
V_{Ga}, Pb	5.777			78C
Se	5.789			
Si	5.839			
S	6.870			
Ge	5.882			
C	5.913			

shallow acceptors, binding energies

(F–B: free to bound transition, DAP: donor- acceptor-pair)

$1S_{3/2}$ calculation	25.82		Effective mass theory using valence band parameters of [76S2] for the calculation	74B 78K
Be	28.0	5	F–B Photoluminescence	75A
C	26.9	≈20	Far infrared photoconductivity	78K
	26.0	5	F–B Photoluminescence	75A
Cd	34.7	5	F–B Photoluminescence	75A
Ge	40.4	5	F–B Photoluminescence	75A
Mg	28.7	≈20	Far infrared photoconductivity	78K
	28.4	5	F–B Photoluminescence	75A
Mn	113.1	28	F–B Photoluminescence	74S2
Si	34.8	≈20	Far infrared photoconductivity	78K
	34.5	5	F–B Photoluminescence	75A
	35.2	5	F–B Photoluminescence	80K2
Sn	170.5	5	F–B Photoluminescence	75A
	167.2	1.8	DAP Photoluminescence	76S2
Zn	30.6	≈20	Far infrared photoconductivity	78K
	30.7	5	F–B Photoluminescence	75A

Copper complexes

Copper is a potentially important contaminant in GaAs. It has been known for many years that copper introduces electrically active centers producing, for example, double compensation [64H, 84B]. These levels of n-type GaAs exhibit thermal activation energies measured from the valence band of 0.156 eV and 0.45 eV. There are also several photoluminescence lines which have been associated with copper or copper-related complexes.

Since copper occupying a substitutional gallium site, Cu_{Ga}, would be expected to act as a double acceptor, many of the electrical and optical signatures of copper-contaminated GaAs have been related to Cu_{Ga} or complexes involving Cu_{Ga} and native defects.

energy levels related to isolated, substitutional transition metal impurities

Impurity	E[eV]	Type	Remarks	T[K]	Ref.
Ti	0.23(1)	a	DLTS and capacitance transient measurements		86H
	−1.00(3)	d	DLTS and capacitance transient measurements		86B1
	+0.60(2)	d	DLTS		86G
V	−0.14	a	DLTS		80M1
Cr	+0.750(20)	a	DLTS (capture cross-section)	320	80M2
	+0.40(3)	d	σ_p^0 absorption photoionization		82U
Mn	+0.1130(5)	a	Luminescence	4	74S2
Fe	+0.49	a	Excitation of luminescence	4.2	83S3
	+0.6	a	σ_p^0 obtained from DLOS	218	82L2
	−0.85	a	σ_n^0 obtained from DLOS	218	82L2
Co	+0.14	a	σ_p^0 photoluminescence excitation photoionization	4.2	86D
Ni	+0.20	1st a	Temperature dependent Hall effect		72B
	−0.40(4)	2nd a	DLTS	0	86B2
	+1.03(3)	2nd a	σ_p^0 photoionization from ODLTS	202	86B2
Ag	+0.230	a	DLTS		82Y
Au	+0.397	a	DLTS		82Y

References for 2.10

60G Goldstein, B.: Phys. Rev. **118** (1960) 1024.
60R Reiss, H., Fuller, C.S.: Semiconductors, Hannay, N.B. (ed.), New York: Reinhold **1960**, p. 230.
61G Goldstein, B.: Phys. Rev. **121** (1961) 1305.
61V Vieland, L.J.: J. Phys. Chem. Solids **21** (1961) 318.
63F Fane, R.W., Goss, A.J.: Solid State Electron. **6** (1963) 383.
64B Boltaks, B.I., Shishiyanu, F.S.: Sov. Phys. Solid State **5** (1964) 1680.
64C1 Casey, H.C., Pearson, G.L.: J. Appl. Phys. **35** (1964) 3401.
64C2 Chang, L.L., Pearson, G.L.: J. Appl. Phys. **35** (1964) 1960.
64H1 Hall, R.N., Racette, J.H.: J. Appl. Phys. **35** (1964) 379.
64H2 Holland, M.G.: Proc. 7th Int. Conf. Phys. Semicond., Paris 1964, Dunod, Paris 1964.
64K1 Kendall, D.L.: Appl. Phys. Lett. **4** (1964) 67.
64K2 Kogan, L.M., Meskin, S.S., Ya Goikhmann, A.: Sov. Phys. Solid State **6** (1964) 882.
64S Sokolov, V.I., Shishiyanu, F.S.: Sov. Phys. Solid State **6** (1964) 265.
65M Moore, R.G.: Bull. Am. Phys. Soc. **10** (1965) 731.
65S1 Seltzer, M.S.: J. Phys. Chem. Solids **26** (1965) 243.
65S2 Straumanis, M.E., Kim. C.D.: Acta. Crystallogr. **19** (1965) 256.
66D Drabble, J.R., Brammer, A.J.: Solid State Commun. **4** (1966) 467.
66L Larrabee, G.B., Osborne, J.F.: J. Electrochem. Soc. **113** (1966) 564.
66P Poltoratskii, E.A., Stuckelbnikov, V.M.: Sov. Phys. Solid State **8** (1966) 770.
68K Kendall, D.L.: Semiconductors and Semimetals, Vol. 4, Willardson, Beer (eds.), New York, London: Academic Press **1968**, p. 163.
69P Panish, M.B., Casey, H.C.: J. Appl. Phys. **40** (1969) 163.
69R Rachmann, J., Biermann, R.: Solid State Commun. **7** (1969) 1771.
69S Showan, S.R., Shaw, D.: Phys. Status Solidi **35** (1969) K79.
71S1 Stringfellow, G.B.: Mater. Res. Bull. **6** (1971) 371.
71S2 Stillman, G.E., Larsen, D.M., Wolfe, C.M., Brandt, R.C.: Solid State Commun. **9** (1971) 2245.
72B Brown, W.J., Blakemore, J.S.: J. Appl. Phys. **43** (1972) 2242.
73A Aspnes, D.E., Studna, A.A.: Phys. Rev. **B7** (1973) 4605.
73C Casey, H.C.: Atomic Diffusion in Semiconductors, Shaw, D. (ed.), New York: Plenum Press **1973**, p. 351.
74B Baldereschi, A., Lipari, N.O.: Phys. Rev. **B9** (1974) 1525.
74S1 Sell, D.D., Casey, H.C., Wecht, K.W.: J. Appl. Phys. **45** (1974) 2650.
74S2 Schairer, W., Schmidt, M.: Phys. Rev. **B10** (1974) 2501.
75A Ashen, D.J., Dean, P.J., Hurle, D.T.J., Mullin, J.B., White, A.M.: J. Phys. Chem. Solids **36** (1975) 1041.
75M1 Mullin, J.B., Straughan, B.W., Driscoll, C.M.H., Willoughby, A.F.W.: CRC Critical Reviews in Solid State Sciences, **1975**, p. 441.

75M2   Mullin, J.B., Straughan, B.W., Driscoll, C.M.H., Willoughby, A.F.W.: Inst. Phys. Conf. Ser. **24** (1975) 275.
75W   Wiley, J.D.: in "Semiconductors and Semimetals", Vol. 10, R.K. Willardson, A.C. Beer eds., Academic Press, New York **1975**.
76C   Chelikowsky, J.R., Cohen, M.L.: Phys. Rev. **B14** (1976) 556.
76H   Hess, K., Bimberg, D., Lipari, N.O., Fischbach, J.K., Altarelli, M.: Proc. 13th Int. Conf. Phys. Semicond., F.G. Fumi ed., Rome **1976**, p. 142.
76S1   Schairer, W., Bimberg, D., Kottler, W., Cho, K., Schmidt, M.: Phys. Rev. **B13** (1976) 3452.
76S2   Skolnick, M.S., Jain, A.K., Stradling, R.A., Leotin, L., Ousset, J.C., Ashkennazy, S.J.: J. Phys. **C9** (1976) 2809.
76T   Tuck, B.: J. Phys. **D9** (1976) 2061.
77B   Borisova, L.A., Akkerman, Z.L., Dorokhov, A.N.: Izv. Akad. Nauk SSSR Neorg. Mater. **13** (1977) 908.
78B   Borisova, L.A., Arkymkhin, P.I., Akkerman, Z.L.: Izv. Akad. Nauk SSSR Neorg. Mater. **14** (1978) 1790.
78C   Cooke, R.A., Hoult, R.A., Kirkman, R.F., Stradling, R.A.: J. Phys. **C11** (1978) 345.
78K   Kirkman, R.F., Stradling, R.A., Lin-Chung, P.J.: J. Phys. **C11** (1978) 419.
78L   Lidow, A., Gibbons, J.F., Deline, V.R., Evans, C.A.: Appl. Phys. Lett. **32** (1978) 572.
78T   Tuck, B., Badawi, M.H.: J. Phys. **D11** (1978) 2541.
79D   Dobson, P.S., Fewster, P.F., Hurle, D.T.J., Hutchinson, P.W., Mullin, J.B., Straughan, B.W., Willoughby, A.F.W.: Inst. Phys. Conf. Ser. **45** (1979) 163.
79T   Tuck, B., Adegboyega, G.A.: J. Phys. **D12** (1979) 1895.
80C   Chiang, T.C., Knapp, J.A., Aano, M., Eastman, D.E.: Phys. Rev. **B21** (1980) 3513.
80F   Fewster, P.F., Willoughby, A.F.W.: J. Cryst. Growth **50** (1980) 648.
80K1   Kasahara, J., Watanabe, N.: Jpn. J. Appl. Phys. **19** (1980) L151.
80K2   Kunzel, H., Ploog, K.: Appl. Phys. Lett. **37** (1980) 416.
80M1   Martin, G.M.: Thesis Univ. P. et M. Curie, **1980**, unpublished.
80M2   Martin, G.M., Mitonneau, A., Pons, D., Mircea, A., Woodard, D.W.: J. Phys. **C13** (1980) 3855.
80N   Nichols, K.H., Yee, C.M.L., Wolfe, C.M.: Solid State Electron. **23** (1980) 109.
80T   Tuck, B., Adegboyega, G.A.: J. Phys. **D13** (1980) 433.
80W   Wilson, R.G., Yasudev, P.K., Jamba, D.M., Evans, C.A., Deline, V.R.: Appl. Phys. Lett. **36** (1980) 215.
81T   Tuck, B., Powell, R.G.: J. Phys. **D14** (1981) 317.
82B1   Blakemore, J.S.: J. Appl. Phys. **53** (1982) R123.
82B2   Baublitz, M., Ruoff, A.L.: J. Appl. Phys. **53** (1982) 6179.
82B3   Blakemore, J.S.: J. Appl. Phys. **53** (1982) 520.
82L1   Lin, L., Lin, Y., Zhong, X., Zhang, Y., Li, H.: J. Cryst. Growth **56** (1982) 344.
82L2   Leyral, P., Litty, F., Bremond, G., Nouailhat, A., Guillot, G.: Semi-Insulating III-V Materials, Evian 1982, Makran-Ebeid, S., Tuck, B. (eds.). Nantwich: Shiva Publishing, **1982**, p. 192.
82M   Mizutami, T., Honda, T., Ishida, S., Kawasaki, Y.: Solid-State Electron. **25** (1982) 885.
82S1   Soma, T., Satoh, J., Matsuo, H.: Solid State Commun. **42** (1982) 889.
82S2   Small, M.B., Potemski, R.M., Reuter, W., Ghez, R.: Appl. Phys. Lett. **41** (1982) 1068.
82U   Ulrici, W.: Phys. Status Solidi (b) **114** (1982) K87.
82Y   Yan, Z.X., Milnes, A.G.: J. Electrochem. Soc. **129** (1982) 1353.
83A   Aspnes, D.E., Studna, A.A.: Phys. Rev. **B27** (1983) 985.
83B   Büttner, H., Pollmann, J.: Physica **177&118** (1983) 278.
83L1   Lindemann, G., Lassnig, R., Seidenbusch, W., Gornik, E.: Phys. Rev. **B28** (1983) 4693.
83L2   Lee, H.J., Look, D.C.: J. Appl. Phys. **54** (1983) 4446.
83P   Palfrey, H.D., Brown, M., Willoughby, A.F.W.: J. Electron. Mater. **12** (1983) 863.
83S1   Sharma, A.C., Ravindra, N.M., Auluck, S., Srivastava, V.K.: Phys. Status Solidi (b) **120** (1983) 715.
83S2   Soma, T., Kagaya, H.-M.: Phys. Status Solidi (b) **118** (1983) 245.
83S3   Shanabrook, B.V., Klein, P.B., Bishop, S.G.: Physica **B116** (1983) 444.
83W2   Winter, J.J., Leupold, H.A., Ross, R.L., Ballato, A.: J. Appl. Phys. **54** (1983) 5176.
84B   Blakemore, J.S., Rahimi, S.: Semicond. Semimet. **20** (1984).
84G   Greiner, M.E., Gibbons, J.F.: Appl. Phys. Lett. **44** (1984) 750.
84P   Patel, C., Parker, T.J., Jamshidi, H., Sherman, W.F.: Phys. Status Solidi (b) **122** (1984) 461.
84S   Skromme, B.J., Stillman, G.E.: Phys. Rev. **B29** (1984) 1982.
85C   Campi, D., Papuzza, C.: J. Appl. Phys. **57** (1985) 1305.
85D   Drouhin, H.-J., Hermann, C., Lampel, G.: Phys. Rev. **B31** (1985) 3859.
85K   Kopylov, A.A.: Solid State Commun. **56** (1985) 1.
85S   Straub, D., Skibowski, M., Himpsel, F.J.: Phys. Rev. **B32** (1985) 5237.
86B1   Brandt, C.D., Hennel, A.M., Pawlowicz, L.M., Wu, Y.T., Bryskiewicz, T., Lagowski, J., Gatos, H.C.: Appl. Phys. Lett. **48** (1986) 1162.
86B2   Brehme, S., Pickenhain, R.: Solid State Commun. **59** (1986) 469.
86D   Deveaud, B., Lambert, B., Auvray, P., Hennel, A.M., Clerjaud, B., Naud, C.: J. Phys. **C19** (1986) 1251.
86G   Guillot, G., Bremond, G., Bencherifa, A., Nouailhat, A., Ulrici, W.: Semi-Insulating III-V Materials, Hakone 1986, Kukimoto, H., Miyazawa, S. (eds.) Ohmsha, Tokyo, **1986**, p. 483.
86H   Hennel, A.M., Brandt, C.D., Wu, Y.T., Bryskiewicz, T., Ko, K., Lagowski, J., Gatos, H.C.: Phys. Rev. **B33** (1986) 7353.
87K   Kavanagh, K.L.: Ph.D. Thesis, University of Cornell **1987**.

Physical property	Numerical value	Experimental conditions	Experimental method, remarks	Ref.

2.11 Gallium antimonide (GaSb)

Electronic properties

band structure: Fig. 1 (Brillouin zone: see section 1.1, Fig. 2)

The *conduction band* is characterized by three sets of minima, the lowest minimum at Γ, slightly higher minima at the L-points at the surface of the Brillouin zone and even higher minima at the X-points. The *valence band* has the structure common to all zincblende semiconductors.

energies of symmetry points of the band structure (relative to the top of the valence band) (in eV):

$E(\Gamma_{6v})$	-12.00	$-11.64(10)$	symmetry symbols in double
$E(\Gamma_{7v})$	-0.76	$-0.82(7)$	$-0.756(15)$	group notation
$E(\Gamma_{8v})$	0		first row: theoretical [76C],
$E(\Gamma_{6c})$	0.86		$0.822(5)$	see also Fig. 1
$E(\Gamma_{7c})$	3.44		$3.191(5)$	second row: angular resolved photo-
$E(\Gamma_{8c})$	3.77		$3.404(10)$	electron spectroscopy [80C]
$E(\Gamma_{8c})$			$7.9(1)$	third row: electroreflectance data
$E(L_{6v})$	-10.17	$-10.06(10)$		at 10 K [76A]
$E(L_{6v})$	-6.25	$-6.60(10)$	
$E(L_{6v})$	-1.45	$-1.55(10)$	$-1.530(10)$
$E(L_{4,5v})$	-1.00	$-1.10(7)$	
$E(L_{6c})$	1.22		$1.095(10)$
$E(L_{4,5c})$	4.43		$4.36(2)$
$E(L_{6c})$	4.59		$4.49(2)$
$E(X_{6v})$	-9.33	$-9.62(10)$	
$E(X_{6v})$	-6.76	$-6.90(10)$	
$E(X_{6v})$	-2.61	$-3.10(7)$	
$E(X_{7v})$	-2.37	-2.86	
$E(X_{6c})$	1.72		
$E(X_{7c})$	1.79		
$E(\Sigma_{3,4v}^{min})$		$-3.64(15)$	
		$-3.90(15)$	

direct energy gap (in eV):

E_{gx}	$0.8099(1)$	2 K	photoluminescence, excitonic gap	73R
$E_{g,dir}(\Gamma_{8v} - \Gamma_{6c})$	$0.8113(2)$	2 K	from value above, assuming an exciton binding energy of 1.4 meV	
	0.822	0 K, extrapol.	electroreflectance, see	81J
	0.75	300 K	Fig. 2	

critical point and spin orbit splitting energies (in eV):

$E_0 + \Delta_0$ $(\Gamma_{7v} - \Gamma_{6c})$	1.569	27 K	modulation spectroscopy	83A2
$E_1(L_{4,5v} - L_{6c})$	2.185			
$E_1 + \Delta_1$ $(L_{6v} - L_{6c})$	2.622			
$E_L(\Gamma_{8v} - L_{6c})$	0.871			

Physical property	Numerical value	Experimental conditions	Experimental method, remarks	Ref.

intra conduction band energies:

L conduction band minima, energy difference to lowest minimum (in meV):
(see also E_L above)

$\Delta E = E(L_{6c})$ $- E(\Gamma_{6c})$	61 82	300 K	Hall effect, magnetoresistance; first value assuming $m_n(L) = 0.226\,m_0$, second value $m_n(L) = 0.43\,m_0$	84K

X conduction band minima, energy difference to lowest minimum (in meV):

$E(X_{6c}) - E(\Gamma_{6c})$	430(2)	10 K	electroreflectance	76A

camel's back structure of conduction band edge:
(see Fig. 2 of section 2.9 for the meaning of the symbols)

Δ	178 meV		estimate from $k \cdot p$ theory using GaP data	85 K
ΔE	25.1 meV			
k_m	0.127 $(2\pi/a)$			
m_t	0.250 m_0			
$m_\parallel$	1.2 m_0			

structure of top of valence band:
By the linear $E(k)$-term in the structure of the valence band of semiconductors with zincblende lattice the spin-degeneracy is lifted. The maxima of the heavy hole bands are shifted in [111]-direction, that of the light

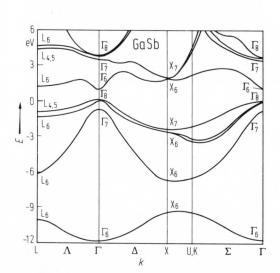

Fig. 1. GaSb. Band structure obtained with a non-local pseudopotential calculation [76C].

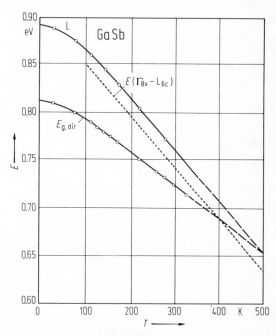

Fig. 2. GaSb. Energies of electroreflectance peaks vs. temperature. $E_{g,dir}$: direct $\Gamma_{8v} - \Gamma_{6c}$ transition, L: indirect $\Gamma_{8v} - L_{6c}$ transition, dotted line: $\Gamma_{8v} - L_{6c}$ gap obtained from L by subtracting the LA phonon energy of 17 meV [81J].

Physical property	Numerical value	Experimental conditions	Experimental method, remarks	Ref.

hole bands in [100]-direction. From an analysis of transport data the following values for the difference of energies at the top of the bands and at $k = 0$ have been found:

$\Delta E[111]$	20 meV			79M
$\Delta E[100]$	5 meV			
$\Delta E[111] -$ $\Delta E[100]$	7.5 meV			85H

effective masses (in units of m_0):

$m_n(\Gamma_{6c})$	0.0412 (12)	1 ··· 30 K	cyclotron resonance of hot electrons; taking into account non-parabolicity and polaron effect a bare effective mass of 0.0396 (21) m_0 results	74H
$m_n(L_{6c})$	0.11 0.95 0.226		transverse longitudinal density of states analysis of transport data	81L
$m_n(X_{6c})$	0.22 1.51		transverse longitudinal; analysis of transport data	81L
$m_{p,h}$ $m_{p,l}$ $m_{p,ds}$	0.28 (13) 0.050 (5) 0.82		conductivity and density of state masses from an analysis of transport data	85H

electron g-factor:

g_c	-7.8 (8)	30 K	stress modulated magneto-reflectance	72R2

valence band parameters:

A	-11.7		calculated using $k \cdot p$ theory	75W
B	-8.19			
$|C|$	11.07			

Lattice properties

Structure

GaSb I	space group $T_d^2 - F\bar{4}3m$ (zincblende structure)		stable at normal pressure	
GaSb II	white tin structure space group $D_{4h}^{19} - I4_1/amd$		high-pressure phase	

transition pressure:

p_{tr}	7.65 (10) GPa	RT	first order Raman scattering	84A

lattice parameter:

a	6.09593 (4) Å	298.15 K	powder, X-ray	65S

temperature dependence: $a = a_0 + a_1 T + a_2 T^2 + a_3 T^3 + a_4 T^4$ with

a_0	6.095882 Å	T in °C	up to 680 °C	
a_1	$3.4963 \cdot 10^{-5}$ Å °C^{-1}			

Physical property	Numerical value	Experimental conditions	Experimental method, remarks	Ref.
a_2	$3.3456 \cdot 10^{-8}$ Å °C^{-2}			
a_3	$-4.6309 \cdot 10^{-11}$ Å °C^{-3}			
a_4	$2.6369 \cdot 10^{-14}$ Å °C^{-4}			

thermal expansion: Fig. 3.

density:

d	$5.6137(4)$ g cm^{-3}	300 K	small variation with temperature (5.60 g cm^{-3} at 900 K)	65S

melting point:

T_m	985 K			69G

phonon dispersion relations: Fig. 4

phonon wavenumbers (in cm^{-1}):
For phonon frequencies obtained from inelastic neutron scattering, see Fig. 4.

$\bar{v}_{TO}(\Gamma)$	$223.6(3)$	RT	first order Raman scattering	84A
$\bar{v}_{LO}(\Gamma)$	$232.6(3)$			

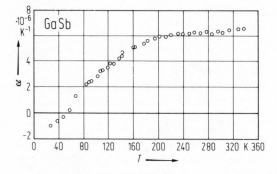

Fig. 3. GaSb. Linear thermal expansion coefficient vs. temperature measured with a quartz dilatometer. High temperature range [63N].

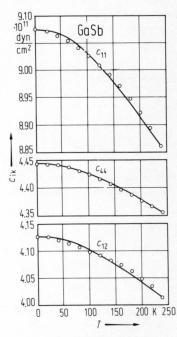

Fig. 5. GaSb. Second order elastic moduli vs. temperature [75B].

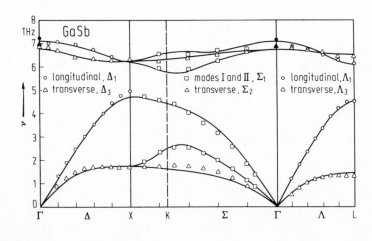

Fig. 4. GaSb. Phonon dispersion relations. Symbols: experimental results from inelastic neutron scattering; full lines: 14 parameter shell model calculation [75F].

Physical property	Numerical value	Experimental conditions	Experimental method, remarks	Ref.
$\bar{v}_{TA}(L)$	46	300 K	second order Raman effect	76K
$\bar{v}_{TA}(X)$	56			
$\bar{v}_{TA}(W)$	75			
$\bar{v}_{LA}(L)$	155			
$\bar{v}_{LO}(L)$	204			
$\bar{v}_{LO}(X)$	210			
$\bar{v}_{TO}(L, X, \Sigma)$	218			

elastic moduli (in 10^{11} dyn cm^{-2}):

c_{11}	8.834	296 K	ultrasound	75B
c_{12}	4.023			
c_{44}	4.322			

Transport properties

Transport in *n-type* GaSb is complicated by the contribution of three sets of conduction bands with minima situated at Γ, L and X. Data on transport coefficients can be consistently explained by a three band model [81L], the X-bands contributing to transport above 180 °C.

In *p-type* GaSb a multiellipsoidal model has to be used at *low temperatures* taking into account the shift of the heavy and light hole band away from $k = 0$. At *high temperatures* a warped sphere model (as in Si and Ge) is adequate. Analyses of transport data between 77 K and RT have to take into account the intermediate region between the two limiting cases.

electron mobility (in cm^2/Vs):

μ_n	3750	Γ-minimum	typical RT values for the contri-	81L
	482	L-minima	butions of the three bands,	
	107	X-minima	obtained from a sample with	
			$n = 1.49 \cdot 10^{18}$ cm^{-3}	

Further data for the L-band: $500 \cdots 800$ cm^2/Vs at RT, $800 \cdots 1600$ cm^2/Vs at 120 K [84K1].
For temperature dependence and the contributions of various scattering mechanisms to the mobility, see Fig. 6.

hole mobility:

μ_p	680 cm^2/Vs	300 K	typical values for Hall	85H
	2500 cm^2/Vs	77 K	mobilities with hole concen-	
			trations of 10^{17} cm^{-3}	

For temperature dependence and the contributions of heavy and light holes, see Fig. 7.

thermal conductivity: Fig. 8.

Optical properties

dielectric constants:

$\varepsilon(0)$	15.69	300 K	reflectance and oscillator fit	62H
$\varepsilon(\infty)$	14.44			

Fig. 7. GaSb. Hall mobility vs. temperature for two samples with acceptor concentrations of $8 \cdot 10^{17}$ cm^{-3} (1) and $3 \cdot 10^{18}$ cm^{-3} (2). (above) Total Hall mobility, (below) contributions from heavy (h) and light (l) holes [79M].

▶

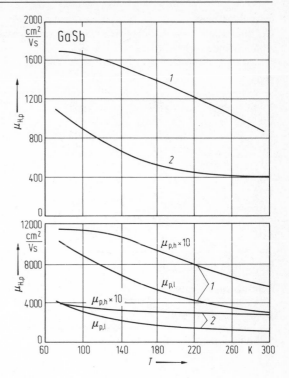

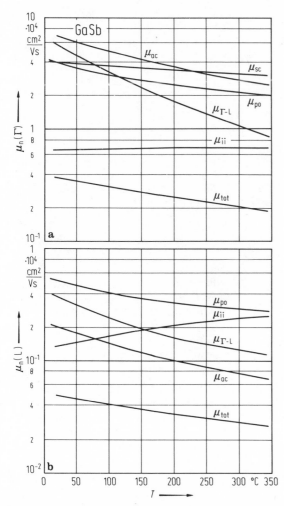

Fig. 6. GaSb. Electron mobility in (a) the Γ-band and (b) the L-band vs. temperature for a sample with $n = 1.49 \cdot 10^{18}$ cm^{-3} (μ_{tot}) and the contributions from various scattering mechanisms (ac – longitudinal acoustic phonon scattering, po – polar optical mode scattering, ii – ionized impurity scattering, sc – space charge scattering, $\Gamma - L$ – intervalley scattering) [81L].

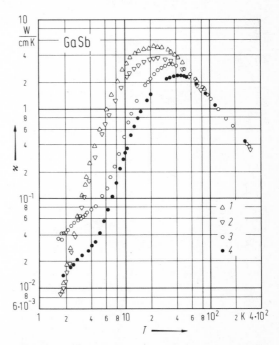

Fig. 8. GaSb. Thermal conductivity of two n-type samples with impurity content of $4 \cdot 10^{18}$ cm^{-3} (1) and $1.4 \cdot 10^{18}$ cm^{-3} (2) and of two p-type samples with $1 \cdot 10^{17}$ cm^{-3} (3) and $2 \cdot 10^{17}$ cm^{-3} (4) [64H].

optical constants

real and imaginary parts of the dielectric constant measured by spectroscopical ellipsometry; n, k, R, K calculated from these data [83A1]; see also Fig. 9.

$\hbar\omega$[eV]	ε_1	ε_2	n	k	R	K[10^3 cm^{-1}]
1.5	19.135	3.023	4.388	0.344	0.398	52.37
2.0	25.545	14.442	5.239	1.378	0.487	279.43
2.5	13.367	19.705	4.312	2.285	0.484	579.07
3.0	9.479	15.738	3.832	2.109	0.444	641.20
3.5	7.852	19.267	3.785	2.545	0.485	902.86
4.0	−1.374	25.138	3.450	3.643	0.583	1477.21
4.5	−8.989	10.763	1.586	3.392	0.651	1547.17
5.0	−5.693	7.529	1.369	2.751	0.585	1394.02
5.5	−5.527	6.410	1.212	2.645	0.592	1474.51
6.0	−4.962	4.520	0.935	2.416	0.610	1469.28

Impurities and defects

diffusion coefficients

Element	D_0[cm^2 s^{-1}]	Q[eV]	T[°C]	Remarks	Ref.
Self-diffusion coefficients					
Ga	$3.2 \cdot 10^3$	3.15	680 ⋯ 700	radiotracer	57E
Sb	$3.4 \cdot 10^4$	3.45	680 ⋯ 700	radiotracer	57E
Impurity diffusion coefficients					
In	$1.2 \cdot 10^{-7}$	0.53		radiotracer	60B
In	dependence on stoichiometry			SIMS	80M
Sn	$2.4 \cdot 10^{-5}$	0.80		radiotracer	60B
Sn	dependent on carrier density			radiotracer	75U
Te	$3.8 \cdot 10^{-4}$	1.20		radiotracer	60B
Cd	$1.5 \cdot 10^{-6}$	0.72		pn-junction depth measurement	68B
Zn	isoconcentration $D = 1.8 \cdot 10^{-11}$ cm^2 s^{-1} at 560 °C			radiotracer (isoconcentration and chemical diffusion)	74D
Zn	$4.0 \cdot 10^{-2}$	1.6		pn-junction	81K

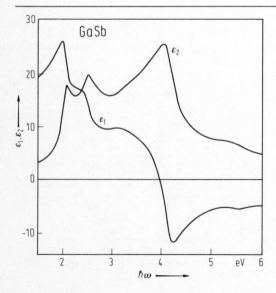

Fig. 9. GaSb. Real and imaginary parts of the dielectric constants vs. photon energy [83A1].

donors

Undoped, relatively pure GaSb is usually p-type. Thus, neutral donor states are not populated in equilibrium. No observations of inter-donor transitions have been reported. Conductivity and Hall coefficient relaxation experiments give evidence of a trap which is related to the S donor [79D].

binding energies of donors

Impurity	E_b [meV]	T [K]	Remarks	Ref.
Te(L)	20(5)	230	pressure dependence of Hall coefficient	69P
Te(X)	<80			71V
Se(L)	85	77	pressure dependence of conductivity	66K
Se(L)*(?)	30	4.2	Shybnikov-de Haas effect	76H
Se(X)	200	290		71V
S(L)	140···150			69P, 71V
S(X)	≈300			71V

acceptors

The dominant acceptor of undoped GaSb seems to be a native defect. Hall measurements show that this acceptor is doubly ionizable [65B, 72J]. The model of a $V_{Ga}Ga_{Sb}$ complex is in agreement with the results of thermodynamical investigations.

binding energies of acceptors

Impurity	E_b [meV]	T [K]	Remarks	Ref.
Si	9.4(5)	1.8	photoluminescence	76J
	13···15	4.2	photoluminescence	71B
Ge	9.5(5)	1.8	photoluminescence	76J
A	34.5	2	unknown impurities, photoluminescence of	72R1
B	54		pair-band. The C-acceptor binds a second	
C	102		hole. For this hole binding energies	72J
D	8		of 34.5 meV and 12 meV are reported.	72R1
E	12			
F	70(5)	12	The E-line is probably caused by an A^+-state	72K
G	130		of the B-acceptor	

References for 2.11

57E   Eisen, F.H., Birchenall, C.E.: Acta Metall. **5** (1957) 265.
60B   Boltaks, B.I., Gutorov, Yu.A.: Sov. Phys. Solid State **1** (1960) 930.
63N   Novikova, S.I., Abrikosov, N.Kh.: Sov. Phys. Solid State (English Transl.) **5** (1963) 1558; Fiz. Tverd. Tela **5** (1963) 2138.
64H   Holland, M.G.: Proc. Int. Conf. Phys. Semicond., Paris 1964, Dunod, Paris 1964 p. 713.
65B   Baxter, R.D., Bate, R.T., Reid, F.J.: J. Phys. Chem. Solids **26** (1965) 14.
65S   Straumanis, M.E., Kim, C.D.: J. Appl. Phys. **36** (1965) 3822.
66K   Kosicki, B.B., Paul. W.: Phys. Rev. Lett. **17** (1966) 246.
68B   Bougnot, J., Szepessy, L., Du Cunha, S.F.: Phys. Status Solidi **26** (1968) K127.
69G   Glazov, V.M., Chizhevskaya, S.N., Evgen'ev, S.B.: Zh. Fiz. Khim. **43** (1969) 373.
69P   Pitt, G.D.: High Temp. – High Pressures **7** (1969) 111.
71B   Burdiyan, I.I., Mal'tsev, S.B., Mironov, I.F., Shreter, Yu.G.: Sov. Phys. Semicond. (English Transl.) **5** (1972) 1734; Fiz. Tekh. Poluprovodn. **5** (1971) 1996.
71V   Vul, A.Ya., Bir, G.L., Shmartsev, Y.V.: Sov. Phys. Semicond. (English Transl.) **4** (1971) 2005; Fiz. Tekh. Poluprovodn. **4** (1970) 2331.
72J   Jakowetz, W., Rühle, W., Breuninger, K., Pilkuhn, M.H.: Phys. Status Solidi (a) **12** (1972) 169.
72K   Kyuregyan, A.S., Lazareva, I.K., Stuchebnikov, V.M., Yunovich, A.E.: Sov. Phys. Semicond. (English Transl.) **6** (1972) 208; Fiz Tekh. Poluprovodn. **6** (1971) 242.
72R1 Rühle, W., Jakowetz, W., Pilkuhn, M.H.: Proc. Int. Conf. Luminescence, Leningrad **1972**, p. 444.

72R2   Reine, M., Aggarwal, R.L., Lax, B.: Phys. Rev. **B5** (1972) 3033.
73R   Rühle, W., Jakowetz, W., Wölk, C., Linnebach, R., Pilkuhn, M.: Phys. Status Solidi (b) **73** (1973) 225.
74H   Hill, D.A., Schwerdtfeger, C.F.: J. Phys. Chem. Solids **35** (1974) 1533.
74D   Du Cunha, S.F., Bougnot, J.: Phys. Status Solidi (a) **22** (1974) 205.
75B   Boyle, W.F., Sladek, R.J.: Phys. Rev. **B11** (1975) 2933.
75F   Farr, M.K., Taylor, J.G., Sinha, S.K.: Phys. Rev. **B11** (1975) 1587.
75U   Uskov, V.A.: Sov. Phys. Semicond. **8** (1975) 1573.
75W   Wiley, J.D.: in "Semiconductors and Semimetals", Vol. 10, R.K. Willardson and A.C. Beer eds., Academic Press, New York and London **1975**.
76A   Aspnes, D.E., Olson, C.G., Lynch, D.W.: Phys. Rev. **B14** (1976) 4450.
76C   Chelikowsky, J.R., Cohen, M.L.: Phys. Rev. **B14** (1976) 556.
76H   Hoo, K., Becker, W.M., Sun, R.-Y.: Solid State Commun. **18** (1976) 313.
76J   Jakowetz, W., Barthruff, D., Benz, K.W.: Proc. 6th Int. Symp. on GaAs and related compounds, Hilsum, C. (ed.), London: Inst. Phys. **1976**, p. 41.
76K   Klein, P.B., Chang, R.K.: Phys. Rev. **B14** (1976) 2498.
79D   Dmoski, L., Baj, M., Kubalski, M., Piotrzkowski, R., Porowski, S.: Inst. Phys. Conf. Ser. **43** (1979) 417.
79M   Mathur, P.C., Jain, S.: Phys. Rev. **B19** (1979) 3152.
80C   Chiang, T.C., Eastman, D.E.: Phys. Rev. **B22** (1980) 2940.
80M   Mathiot, D., Edelin, G.: Philos. Mag. **A41** (1980) 447.
81J   Joullié, A., Zein Eddine, A., Girault, B.: Phys. Rev. **B23** (1981) 928.
81K   Kagawa, T., Motosugi, G.: Jpn. J. Appl. Phys. **20** (1981) 597.
81L   Lee, H.J., Woolley, J.C.: Can. J. Phys. **59** (1981) 1844.
83A1   Aspnes, D.E., Studna, A.A.: Phys. Rev. **B27** (1983) 985.
83A2   Alibert, C., Joullié, A., Joullié, A.M., Ance, C.: Phys. Rev. **B27** (1983) 4946.
84A   Aoki, K., Anastassakis, E., Cardona, M.: Phys. Rev. **B30** (1984) 681.
84K   Kourkoutas, C.D., Bekris, P.D., Papaioannou, G.J., Euthymiou, P.C.: Solid State Commun. **49** (1984) 1071.
85H   Heller, M.W., Hamerly, R.G.: J. Appl. Phys. **57** (1985) 4626.
85K   Kopylov, A.A.: Solid State Commun. **56** (1985) 1.

Physical property	Numerical value	Experimental conditions	Experimental method, remarks	Ref.

2.12 Indium nitride (InN)

Electronic properties

The band structure (Fig. 1, Brillouin zone: Fig. 3 of section 2.1) shows a direct gap at Γ, closely similar to that of GaN.

energy gap (in eV):

$E_{g,dir}$	2.05(1)	300 K	absorption edge,	77T
	2.11	78 K		75O
	1.89	RT		86F

effective mass:

m_{opt}	$0.11 m_0$	$T = 300$ K	plasma edge	77T

Calculated effective masses of $m_n = 0.12 m_0$, $m_{p,h} = 0.5 m_0$, $m_{p,l} = 0.17 m_0$ are reported [86F].

Lattice properties

lattice parameter:

a	3.5446 Å		epitaxial layers, X-ray	78P
c	5.7034 Å			

density:

d	6.81 g cm^{-3}	298.15 K	X-ray	67P

Physical property	Numerical value	Experimental conditions	Experimental method, remarks	Ref.
melting point:				
T_m	1373 K			70M
phonon wavenumbers:				
$\bar{v}_\mathrm{TO}(\Gamma)$	478 cm^{-1}	$T = 300$ K	reflectivity, Kramers–Kronig analysis	75O
$\bar{v}_\mathrm{LO}(\Gamma)$	694 cm^{-1}			

Transport properties

electrical conductivity:

σ	$2\cdots3\cdot10^2\,\Omega^{-1}\,\mathrm{cm}^{-1}$	$T = 300$ K	temperature coefficient of resistivity $3.7\cdot10^{-3}\,\mathrm{K}^{-1}$ at $200\cdots300$ K for pressed powder	56J, 72H, 74T

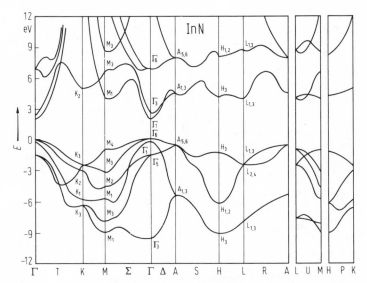

Fig. 1. InN. Band structure calculated by a pseudopotential method [86F].

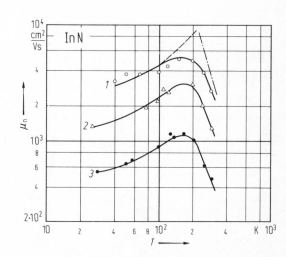

Fig. 2. InN. Electron mobility vs. temperature for three samples with RT carrier concentrations of $5.3\cdot10^{16}$ (1), $7.5\cdot10^{16}$ (2) and $1.8\cdot10^{17}\,\mathrm{cm}^{-3}$ (3). Broken line: calculated ionized impurity scattering mobility, dot-dashed line: empirical high-temperature mobility ($\mu \propto T^{-3}$) for sample 1. Solid lines: total mobility calculated for each sample [84T].

Physical property	Numerical value	Experimental conditions	Experimental method, remarks	Ref.
electron mobility:				
μ_n	$250\,(50)\,\text{cm}^2/\text{Vs}$	$T = 300\,\text{K}$	see also Fig. 2	72H
	$20\,\text{cm}^2/\text{Vs}$	$300\,\text{K}$		74T
	$35 \cdots 50\,\text{cm}^2/\text{Vs}$	$300\,\text{K}$		77M

Optical properties

refractive index:

n	2.56	$T = 300\,\text{K}$, $n = 3 \cdot 10^{20}\,\text{cm}^{-3}$ $\lambda = 1.0\,\mu\text{m}$	interference method	77T
	2.93	$0.82\,\mu\text{m}$		
	3.12	$0.66\,\mu\text{m}$		

dielectric constant:

$\varepsilon(\infty)$	9.3	heavily doped film	infrared reflectivity	77T

References for 2.12

56J   Juza, R., Rabenau, A.: Z. Anorg. Allg. Chem. **285** (1956) 212.
67P   Pearson, W.B.: A Handbook of Lattice Spacings and Structures of Metals and Alloys, Pergamon Press, Oxford-London 1967.
70M   MacChesney, J.B., Bridenbaugh, P.M., O'Connor, P.B.: Mater. Res. Bull. **5** (1970) 783.
72H   Hovel, H.J., Cuomo, J.J.: Appl. Phys. Lett. **20** (1972) 71.
74T   Trainor, J.W., Rose, K.: J. Electron. Mater. **3** (1974) 821.
75O   Osamura, K., Naka, S., Murakami, Y.: J. Appl. Phys. **46** (1975) 3432.
77M   Marasina, L.A., Pichugin, E.G., Tlaczala, M.: Krist. Techn. **12** (1977) 541.
77T   Tyagai, V.A., Evstigneev, A.M., Krasiko, A.N., Adreeva, A.F., Malakhov, V.Ya.: Sov. Phys. Semicond. (English Transl.) **11** (1977) 1257; Fiz. Tekh. Poluprovodn. **11** (1977) 2142.
78P   Pichugin, I.G., Tlachala, M.: Izv. Akad. Nauk SSSR, Neorg. Mater. **14** (1978) 175.
84T   Tansley, T.L., Foley, C.P.: Electron. Lett. **20** (1984) 1087.
86F   Foley, C.P., Tansley, T.L.: Phys. Rev. **B33** (1986) 1430.

2.13 Indium phosphide (InP)

Electronic properties

band structure: Fig. 1 (Brillouin zone: see section 1.1, Fig. 2)

InP is a direct semiconductor. The *conduction band* minimum is situated at Γ. Higher conduction band minima at L and X have been detected in optical experiments. The X band minima show no camel's back structure, in contrast to most other III–V compounds with zincblende structure [85K]. The *valence band* has the structure common to all zincblende type semiconductors.

energies at symmetry points of the band structure (relative to the top of the valence band) (in eV):

$E(\Gamma_{6v})$	-11.42	$-0.108\,(\text{e})$	symmetry symbols in double
$E(\Gamma_{7v})$	-0.21		group notation
$E(\Gamma_{8v})$	0		first row:
$E(\Gamma_{6c})$	1.50	$1.42\,(\text{e})$	calculated data from [76C],
$E(\Gamma_{7c})$	4.64	$4.8\,(\text{b})$	see Fig. 1
$E(\Gamma_{8c})$	4.92	$4.87\,(\text{a})$	
$E(X_{6v})$	-8.91		
$E(X_{6v})$	-6.01	$-5.9\,(\text{f})$	

Physical property	Numerical value	Experimental conditions	Experimental method, remarks	Ref.
$E(E_{6v})$	-2.09		second row:	
$E(X_{7v})$	-2.06	-2.2 (f)	comparison with experiment	
$E(X_{6c})$	2.44	2.38 (d, e)	using data on critical points	
$E(X_{7c})$	2.97		(a:[66S], b:[68M], c:[70J],	
$E(L_{6v})$	-9.67		d:[72O], e:[80C2], f:[83W])	
$E(L_{6v})$	-5.84			
$E(L_{6v})$	-1.09	-1.23 (b)		
$E(L_{4,5v})$	-0.94	-1.12 (b)		
$E(L_{6c})$	2.19	2.03 (c, e)		
$E(L_{6c})$	5.58			
$E(L_{4,5c})$	5.70			

direct energy gap:

$E_{g,dir}$ $(\Gamma_{8v} - \Gamma_{6c})$	$1.4236(1)$ eV	1.6 K	from excitonic gap (below)	85M
	1.344 eV	300 K	absorption, photoluminescence	85B

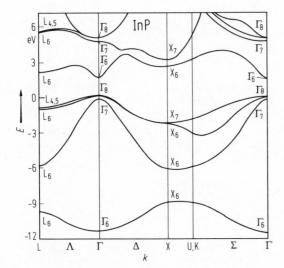

Fig. 1. InP. Band structure obtained with a non-local pseudopotential method [76C].

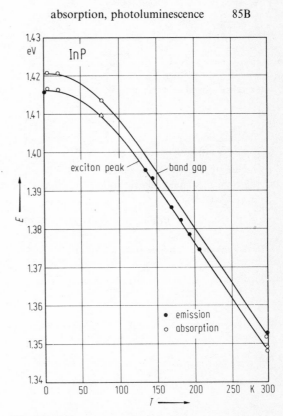

Fig. 2. InP. Energy gap and exciton peak energy vs. temperature below RT from absorption and emission data [64T].

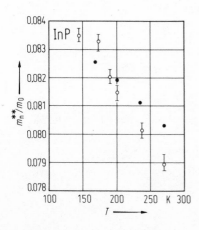

◄ Fig. 3. InP. Polaronic effective mass vs. temperature. Full circles: [71E], open circles: [84M].

Physical property	Numerical value	Experimental conditions	Experimental method, remarks	Ref.

temperature dependence of direct energy gap (in $10^{-4}\,\text{eV K}^{-1}$):

$dE_{\text{g,dir}}/dT$	-2.9	below RT	absorption, Fig. 2	64T
	-5.1	above RT	luminescence	83K

exciton energies (obtained from fitting of exciton fine structure in reflectivity at $T = 1.6\,\text{K}$ [85M]):

$E(1S, T)$	$1.41850(5)\,\text{eV}$	1.6 K	transverse exciton, ground state	85M
$E(1S, L)$	$1.41867(7)\,\text{eV}$		longitudinal exciton, ground state	
E_b	$5.1(1)\,\text{meV}$	1.6 K	exciton binding energy	
Δ_{ex}	$0.04(2)\,\text{meV}$		exchange interaction energy	

intra conduction band energies (in eV):

$E(X_{6c} - \Gamma_{6c})$	$0.960(5)$	8 K	wavelength derivative transmission	72O
$E(L_{6c} - \Gamma_{6c})$	0.39	RT	capacitance	80W1
	0.61	$0\cdots300\,\text{K}$	photoemission	70J
$\Delta_0'(\Gamma_{8c} - \Gamma_{7c})$	0.07	300 K	electroreflectance	66S

intra valence band energies (in eV):

$\Delta_0(\Gamma_{8v} - \Gamma_{7v})$	$0.108(1)$	5 K	wavelength modulated reflectivity	80C
$\Delta_1(\Lambda_{4,5v} - \Lambda_{6c})$	$0.133(4)$	300 K	ellipsometry	82K

structure of top of valence band:
By the linear term in $E(k)$ the maxima of the heavy hole valence band are shifted on the [111] axes, the energy of the band maximum relative to the energy at $k = 0$ being:

$\Delta E[111]$	6 meV	78 K	electroabsorption	77V
	38 meV	298 K		

effective masses, electrons (in units of m_0):

m_n	$0.0765(5)$	4.2 K	bare band edge mass obtained from polaron cyclotron resonance converted from polaron mass using a Fröhlich coupling constant $\alpha = 0.15(1)$	85H
m_n^{**}	$0.079(0.075)$	270 K	polaron (bare) band edge mass from magnetophonon resonance, see Fig. 3 for temperature dependence	84M
	$0.084(0.080)$	143 K		

electron g-factor:

g_c	$1.48(5)$	4.2K	modulated photovoltaic effect	75R

effective masses, holes (in units of m_0):

$m_{\text{p,h}}$	$0.60(2)$	$110\,\text{K}, B \parallel [111]$	cyclotron resonance	74L
	$0.56(2)$	$B \parallel [100]$		
$m_{\text{p,l}}$	$0.12(1)$			
m_{so}	$0.121(1)$			

Physical property	Numerical value	Experimental conditions	Experimental method, remarks	Ref.
valence band parameters:				
A	-6.28		calculated using $k \cdot p$ theory	75W
B	-4.17			
$|C|$	6.24			

Lattice properties

Structure

InP I	space group $T_d^2 - F\bar{4}3m$ (zincblende structure)		stable at normal pressure	
InP II	space group $O_h^5 - Fm3m$ $a = 5.514\,\text{Å}$		high-pressure phase	63J

transition pressure:

p_{tr}	133 kbar			63J

lattice parameter:

a	5.8687(10) Å	291.15 K	powder, X-ray	58G
	5.8871(6) Å	912 K		72K

linear thermal expansion coefficient: Fig. 4.

density:

d	4.81 g cm^{-3}			77M

melting point:

T_m	1335(5) K		from a critical analysis of literature data	84T

phonon dispersion relations: Fig. 5.

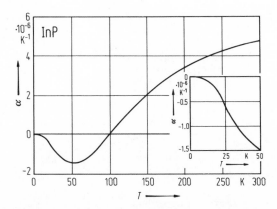

Fig. 4. InP. Linear thermal expansion coefficient vs. temperature calculated using experimental pressure derivatives of elastic moduli and phonon wavenumbers [82S2].

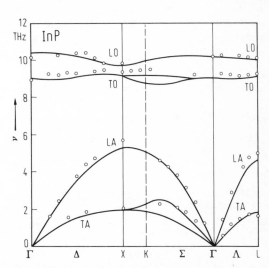

Fig. 5. InP. Phonon dispersion relations calculated by an eight parameter bond-bending force model (solid lines) [80K] and experimental data from inelastic neutron scattering [75B].

Physical property	Numerical value	Experimental conditions	Experimental method, remarks	Ref.
phonon frequencies (in THz):				
$\nu_{LO}(111)(0.05)$	10.3 (3)	RT, $n = 10^{12}\,cm^{-3}$	coherent inelastic neutron scattering	78B
$\nu_{TO}(\Gamma_{15})$	9.2 (2)			
$\nu_{TA}(X_5)$	2.05 (10)			
$\nu_{LA}(X_3)$	5.8 (3)			
$\nu_{TO}(X_5)$	9.70 (10)			
$\nu_{LO}(X_1)$	9.95 (20)			
$\nu_{TA}(L_3)$	1.65 (2)			
$\nu_{LA}(L_1)$	5.00 (10)			
$\nu_{TO}(L_3)$	9.50 (15)			
$\nu_{LO}(L_1)$	10.2 (3)			
$\nu_{LO}(\Gamma)$	10.32 (344.5 cm^{-1})		Raman scattering, zone center phonons	80T
$\nu_{TO}(\Gamma)$	9.093 (303.3 cm^{-1})			
second order elastic moduli (in 10^{11} dyn cm^{-2}):				
c_{11}	10.11	RT	ultrasonic wave transit times	80N
c_{12}	5.61			
c_{44}	4.56			
third order elastic moduli (in 10^{12} dyn cm^{-2}):				
c_{111}	−8.6		calculated from elastic moduli, their pressure dependence and other literature parameters	80N
c_{112}	−1.85			
c_{123}	−5.1			
c_{144}	−6.5			
c_{166}	+1.6			
c_{456}	−0.042			

Transport properties

The transport in n-type samples is determined up to 800 K by the electrons in the Γ_{6c}-minimum. Only at very high temperatures L_{6c}-electrons contribute to the conductivity. In high purity samples polar scattering dominates at room temperature.

intrinsic carrier concentration (in cm^{-3}):

n_i	$1.2 \cdot 10^8$	293 K	from an analysis of transport data	56W
	$8.4 \cdot 10^{15}\, T^{3/2} \exp(-1.34/2\,kT)$		for the range 700⋯920 K (T in K, kT eV) see Fig. 6	55F

electron mobility (typical data from the literature, in cm^2/Vs):

$\mu_{H,n}$	131 600	77 K, VPE layer, $n = 2.5 \cdot 10^{14}\,cm^{-3}$		85Z
	5370	300 K		
	5900	300 K	theoretical upper limit	80W2

Mobility values for single crystal and polycrystalline samples are mostly lower. Best values at 300 K are around 4000 cm^2/Vs. At 77 K values up to 130000 cm^2/Vs have been reported in [82A].

Physical property	Numerical value	Experimental conditions	Experimental method, remarks	Ref.

Temperature dependence of electron mobility: Fig. 7.

hole mobility:

$\mu_{H,p}$	$150\,(T/300\,\text{K})^{-2.2}\,\text{cm}^2/\text{Vs}$		above 200 K; temperature dependence, see Fig. 8	75W

thermal conductivity: Fig. 9.

Optical properties

dielectric constants:

$\varepsilon(0)$	12.5(1)		fit of transport data	80L
	12.56(20)	300 K	capacitance measurements	86M
	11.93(20)	77 K		
	12.61		oscillator fit of infrared	62H
$\varepsilon(\infty)$	9.61		reflectance	

According to [86M] the temperature dependence of $\varepsilon(0)$ below 220 K is given by $\varepsilon(0) = 11.76(1 + 2.26 \cdot 10^{-4}\,T)$ (T in K).

refractive index: in addition to the data above, see Fig. 10.

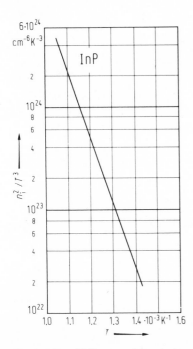

Fig. 6. InP. n_i^2/T^3 vs. reciprocal temperature [55F]. n_i: intrinsic concentration.

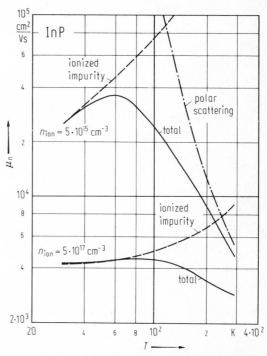

Fig. 7. InP. Electron mobility vs. temperature calculated for two concentrations of ionized impurities. Contributions from scattering mechanisms are indicated (dashed lines) [80W2].

optical constants:

real and imaginary parts of the dielectric constant measured by spectroscopical ellipsometry; n, k, R, K calculated from these data [83A]; see also Fig. 11.

$\hbar\omega$ [eV]	ε_1	ε_2	n	k	R	$K[10^3\,\mathrm{cm}^{-1}]$
1.5	11.904	1.400	3.456	0.203	0.305	30.79
2.0	12.493	2.252	3.549	0.317	0.317	64.32
2.5	14.313	3.904	3.818	0.511	0.349	129.56
3.0	17.759	10.962	4.395	1.247	0.427	379.23
3.5	6.400	12.443	3.193	1.948	0.403	691.21
4.0	6.874	10.871	3.141	1.730	0.376	701.54
4.5	8.891	16.161	3.697	2.186	0.449	996.95
5.0	−7.678	14.896	2.131	3.495	0.613	1771.52
5.5	−4.528	7.308	1.426	2.562	0.542	1428.14
6.0	−2.681	5.644	1.336	2.113	0.461	1285.10

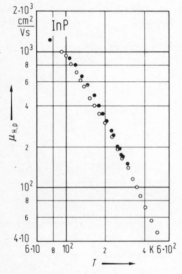

Fig. 8. InP. Hole Hall mobility vs. temperature for pure p-type samples, after [75W].

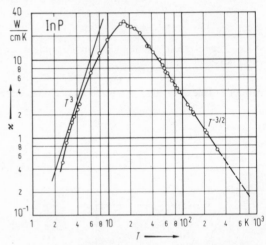

Fig. 9. InP. Thermal conductivity vs. temperature [65A].

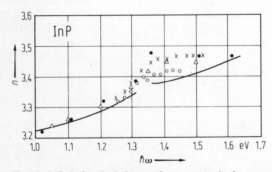

Fig. 10. InP. Refractive index vs. photon energy in the range $1.0\cdots1.6$ eV. Theoretical curve and experimental points from [82B] (triangles), [82S1] (full circles) and [85C] (open circles: p-type, crosses: n-type samples).

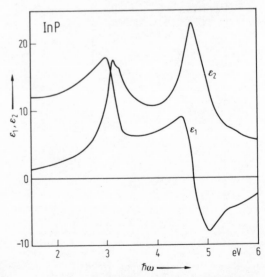

Fig. 11. InP. Real and imaginary parts of the dielectric constant vs. photon energy [83A].

Impurities and defects

diffusion coefficients

Element	$D_0\,[\mathrm{cm^2\,s^{-1}}]$	$Q\,[\mathrm{eV}]$	$T\,[^\circ\mathrm{C}]$	Remarks	Ref.
Self-diffusion coefficients					
In	$1\cdot10^5$	3.85	$838\cdots980$	radiotracer	61G
P	$7\cdot10^{10}$	5.65	$903\cdots1010$	radiotracer	61G
Impurity diffusion coefficients					
Ag	$3.6\cdot10^{-4}$	0.59		radiotracer	69A1
Au	$1.32\cdot10^{-5}$	0.48		radiotracer	69R
Cd	1.8	1.9		radiotracer	67A
Cu	$3.8\cdot10^{-3}$	0.69		radiotracer	69A2
Zn	complex profiles				64C

donors: Discrimination of donor chemical species is made difficult because of the small electron effective mass ($m_c^* < 0.1\,m_0$) and large dielectric constant ($\varepsilon_s \approx 12$). This leads to small values of electron binding energy at donor states. The effective mass donor Rydberg for InP is 7.31 meV [74H1]. The large extend of the electron wave function makes the central cell corrections small ($\psi^2(r=0)$), typically ≈ 0.1 meV. Random electric fields, strain and other effects can produce line widths for donor-related optical transitions which are larger than the central cell corrections, especially at zero magnetic field.

acceptors: Zinc and carbon are often common residual acceptors in InP. The group IV elements, except for C and Ge, do not substitute for P, and so show no amphoteric behavior.

binding energy of acceptors

Element	$E_b\,[\mathrm{meV}]$	$T\,[\mathrm{K}]$	Remarks	Ref.
Zn	48	4.2	Free-bound photoluminescence.	79D
	46.4 (10)	10		76H
Cd	57.0 (10)	1.8	No lineshape fits of emission lines were	72W
Hg	98 (2)	6	attempted introducing some uncertainty	73W
C_p	41.3 (5)	1.8	for the binding energies thus determined.	74H2
	41.5	4.2		79D
Ge_p	210 (20)	2		76W
Mn	270 (10)	6		73W
Cu	$60\cdots73$	6	Probably Cu-vacancy complex	73W
Mg	31	77	Crystal grown by molecular beam epitaxy	77B
	108		Crystal grown by LEC	
Be	31		Crystal grown by molecular beam epitaxy	77B
	143		Crystal grown by gradient free growth	

transition metal impurities

Energy levels related to isolated, substitutional transition metal impurities. ("+" above valence band, "−" below conduction band.)

Impurity	E [eV]	Type	Remarks	T [K]	Ref.
Ti	−0.63(3)	d	DLTS		86B
V	+0.21	d	DLTS + σ_p^0 photoluminescence excitation photoionization	4	86D
Cr	−0.39(1)	a	Temperature dependent resistivity and Hall effect measurements		79I
	+0.96(1)	a	PICTS		82R
	+0.56	d	Temperature dependent Hall measurements		86L
Mn	+0.21	a	Temperature dependent resistivity and Hall measurements		84K
Fe	−0.65	a	Photoconductivity		81E
	+0.7850	a	σ_p^0 absorption photoionization; $^5T_2$ excited state located at +1.1379	1.3	86J
Co	+0.24	a	DLTS and ODLTS		84R
Ni	+0.48(4)	1st a	DLTS		90K
	−0.27(2)	2nd a	DLTS		89K
Au	−0.55	d	DLTS		87P

References for 2.13

55F   Folberth, O.G., Weiss, H.: Z. Naturforsch. **10a** (1955) 615.
56W   Weiss, H.: Z. Naturforsch. **11a** (1956) 430.
58G   Giesecke, G., Pfister, H.: Acta Crystallogr. **11** (1958) 369.
62H   Hass, M., Henvis, B.W.: J. Phys. Chem. Solids **23** (1962) 1099.
63J   Jamieson, J.C.: Science **139** (1983) 845.
61G   Goldstein, B.: Phys. Rev. **121** (1961) 1305.
64C   Chang, L.L., Casey, H.C.: Solid State Electron. **7** (1964) 481.
64T   Turner, W.J., Reese, W.E., Pettit, G.D.: Phys. Rev. **136** (1964) A1467.
65A   Aliev, S.A., Nashelskii, A.Ya., Shalyt, S.S.: Sov. Phys. Solid State (English Transl.) **7** (1965) 1287; Fiz. Tverd. Tela **7** (1965) 1590.
66S   Shaklee, K.L., Cardona, M., Pollak, F.H.: Phys. Rev. Lett. **16** (1966) 48.
67A   Arseni, K.A., Boltaks, B.I., Gordin, V.L., Ugai, Ya.A.: Inorg. Mater. (USSR) **3** (1967) 1465.
68M   Matatagui, E., Thompson, A.E., Cardona, M.: Phys. Rev. **176** (1968) 950.
69A1 Arseni, K.A., Boltaks, B.I.: Sov. Phys. Solid State **10** (1969) 2190.
69A2 Arseni, K.A.: Sov. Phys. Solid State **10** (1969) 2263.
69R   Rembeza, S.I.: Sov. Phys. Semicond. **3** (1969) 519.
70J   James, L.W., van Dyke, J.P., Herman, F., Chang, D.M.: Phys. Rev. **B1** (1970) 3998.
71E   Eaves, L., Stradling, R.A., Askenazy, S., Leotin, J., Portal, J.C., Ulmet, J.P.: J. Phys. **C4** (1971) L42.
72K   Kudman, I., Pfaff, R.J.: J. Appl. Phys. **43** (1972) 3760.
72O   Onton, A., Chicotka, R.J., Yacoby, Y.: Proc. 11th Int. Conf. Phys. Semicond., Warsaw 1972, Polish Scientific Publishers, Warsaw **1972**, p. 1023.
72W   White, A.M., Dean, P.J., Taylor, L.L., Clarke, R.C., Ashen, D.J., Mullin, J.B.: J. Phys. **C5** (1972) 1727.
73W   Williams, E.W., Elder, W., Astles, M.G., Webb, M., Mullin, J.B., Straughan, B., Tulton, P.J.: J. Electrochem. Soc. **120** (1973) 1741.
74H1 Hoult, R.A., Stradling, R.A., Bradley, C.C.: J. Phys. **C7** (1974) 1164.
74H2 Hess, K., Stath, N., Benz, K.W.: J. Electrochem. Soc. **121** (1974) 1208.
74L   Leotin, J., Barbaste, R., Askenazy, S., Skolnick, M.S., Stradling, R.A., Tuchendler, J.: Solid State Commun. **15** (1974) 693.
75B   Borcherds, P.H., Alfrey, G.F., Saunderson, D.H., Woods, A.D.B.: J. Phys. **C8** (1975) 2022.
75R   Rochon, P., Fortin, E.: Phys. Rev. **B12** (1975) 5803.
75W   Wiley, J.D.: in "Semiconductors and Semimetals", Vol. 10, R.K. Willardson, A.C. Beer eds., Academic Press, New York **1975**.
76C   Chelikowsky, J.R., Cohen, M.L.: Phys. Rev. **B14** (1976) 556.
76H   Hess, K.: Dissertation, Stuttgart **1976**.
76W   White, A.M., Dean, P.J., Day, B.: Proc. XIIIth Int. Conf. on Physics of Semiconductors, Fumi, F.G. (ed.), Rome: Tipografia Marves **1976**, p. 1037.
77B   Bimberg, D., Hess, K., Lipari, N.O., Fischebach, J.U., Altarelli, M.: Physica **B89** (1977) 139.
77M   Merrill, L.: J. Phys. Chem. Ref. Data **6** (1977) 1205.
77V   Vorobeev, L.E., Shturbin, A.V., Osokin, F.I.: Sov. Phys. Semicond. (English Transl.) **11** (1977) 879; Fiz. Tekh. Poluprovodn. **11** (1977) 1497.

79D	Dean, P.J., Robbins, D.J., Bishop, S.G.: J. Phys. **C12** (1979) 5567.
79I	Iseler, G.W.: Inst. Phys. Conf. Ser. **45** (1979) 144.
80C	Camassel, J., Merle, P., Bayo, L., Mathieu, H.: Phys. Rev. **B22** (1980) 2020.
80K	Kushwaha, M.S., Kushwaha, S.S.: Canad. J. Phys. **58** (1980) 351.
80L	Lee, H.J., Basinski, J., Juravel, L.Y., Woolley, J.C.: Canad. J. Phys. **58** (1980) 923.
80N	Nichols, D.N., Rimai, D.S., Sladek, R.J.: Solid State Commun. **36** (1980) 667.
80T	Trommer, R., Müller, H., Cardona, M.: Phys. Rev. **B21** (1980) 4869.
80W1	Wada, O., Majerfeld, A., Choudhury, A.N.M.M.: J. Appl. Phys. **51** (1980) 423.
80W2	Walukiewicz, W., Lagowski, J., Jastrzebski, L., Rava, P., Lichtensteiger, M., Gatos, C.H., Gatos, H.C.: J. Appl. Phys. **51** (1980) 2659.
81E	Eaves, L., Smith A.W., Williams, P.J., Cockayne, B., MacEwan, W.R.: J. Phys. **C14** (1981) 5063.
82A	Adamski, J.A.: J. Cryst. Growth **60** (1982) 141.
82B	Burkhard, H., Dinges, H.W., Kuphal, E.: J. Appl. Phys. **53** (1982) 655.
82K	Kelso, S.M., Aspnes, D.E., Pollack, M.A., Nahory, R.E.: Phys. Rev. **B26** (1982) 6669.
82R	Rhee, J.K., Battacharya, P.K.: J. Appl. Phys. **53** (1982) 4247.
82S1	Skolnick, M.S., Dean, P.J.: J. Phys. **C15** (1982) 5863.
82S2	Soma, T., Satoh, J., Matsuo, H.: Solid State Commun. **42** (1982) 889.
83A	Aspnes, D.E., Studna, A.A.: Phys. Rev. **B27** (1983) 985.
83K	Kirillov, D., Merz, J.L.: J. Appl. Phys. **54** (1983) 4104.
83W	Williams, G.P., Cerrina, F., Anderson, J., Lapeyre, G.J., Smith, R.J., Hermanson, J., Knapp, J.A.: Physica **117B & 118B** (1983) 350.
84M	Maeda, Y., Taki, H., Sakata, M., Ohta, E., Yamada, S., Fukui, T., Miura, N.: J. Phys. Soc. Jpn. **53** (1984) 3553.
84K	Kuznetsov, V.P., Messerer, M.A., Omel'yanovskii, E.M.: Fiz. Tekh. Poluprovodn. **18** (1984) 446; Sov. Phys. Semicond. (English Transl.) **18** (1984) 278.
84R	Rojo, P., Leyral, P., Nouailhat, A., Guillot, G., Lambert, B., Deveaud, B., Coquille, R.: J. Appl. Phys. **55** (1984) 395.
84T	Tmar, M., Gabriel, A., Chatillon, C., Ansara, I.: J. Cryst. Growth **68** (1984) 557.
85B	Bugajski, M., Lewandowski, W.: J. Appl. Phys. **57** (1985) 521.
85C	Campi, D., Papuzza, C.: J. Appl. Phys. **57** (1985) 1305.
85H	Helm, M., Knap, W., Seidenbusch, W., Lassnig, R., Gornik, E.: Solid State Commun. **53** (1985) 547.
85K	Kopylov, A.A.: Solid State Commun. **56** (1985) 1.
85M	Mathieu, H., Chen, Y., Camassel, J., Allegre, J., Robertson, D.S.: Phys. Rev. **B32** (1985) 4042.
85Z	Zhu, L.D., Chan, K.T., Ballantyne, J.M.: Appl. Phys. Lett. **47** (1985) 47.
86B	Brandt, C.D., Hennel, A.M., Pawlowicz, L.M., Wu, Y.T., Bryskiewicz, T., Lagowski, J., Gatos, H.C.: Appl. Phys. Lett. **48** (1986) 1162.
86D	Deveaud, B., Plot, B., Lambert, B., Bremond, G., Guillot, G., Nouailhat, A., Clerjaud, B., Naud, C.: J. Appl. Phys. **59** (1986) 3126.
86J	Juhl, A., Bimberg, D.: Semi-Insulating III–V Materials, Hakone 1986, Kukimoto, H., Miyazawa, S. (eds.), OHM, North-Holland, **1986**, p. 477.
86L	Lambert, B., Toudic, Y., Coquille, R., Grandpierre, G., Gauneau, M.: Defects in Semiconductors, Proc. 14th Internat. Conf. Defects in Semicond., Paris (1986), von Bardeleben, H.J. (ed.), Materials Science Forum 10···12, Trans. Tech. Publications, Switzerland, **1986**, p. 651.
86M	Meiners, L.G.: J. Appl. Phys. **59** (1986) 1611.
87P	Parguel, V., Favennec, P.N., Gauneau, M., Rihet, Y., Chaplain, R., L'Haridon, H., Vaudry, C.: J. Appl. Phys. **62** (1987) 824.
89K	Korona, K., Hennel, A.M.: Appl. Phys. Lett. **55** (1989) 1085.
90K	Korona, K., Karpinska, K., Babinski, A., Hennel, A.M.: Acta Phys. Pol. **A77** (1990) 71.

Physical property	Numerical value	Experimental conditions	Experimental method, remarks	Ref.

2.14 Indium arsenide (InAs)

Electronic properties

band structure: Fig. 1 (Brillouin zone: see section 1.1, Fig. 2)
InAs resembles in its band structure InSb, having only a slightly larger energy gap and a smaller spin-orbit splitting of the top of the valence band.

energies of symmetry points of the band structure (relative to the top of the valence band) (in eV):

$E(\Gamma_{6v})$	−12.69	−12.3(4)	symmetry symbols in double
$E(\Gamma_{7v})$	−0.43		group notation

Physical property	Numerical value	Experimental conditions	Experimental method, remarks	Ref.
$E(\Gamma_{8v})$	0.00		first row: theoretical [76C]	
$E(\Gamma_{6c})$	0.37		second row: XPS [74L] and	
$E(\Gamma_{7c})$	4.39		angular resolved photo-	
$E(\Gamma_{8c})$	4.63		emission [83W] data (spin-	
$E(X_{6v})$	-10.20	$-9.8(3)$	splitting not resolved)	
$E(X_{6v})$	-6.64	$-6.3(2)$		
$E(X_{6v})$	-2.47	$-2.4(3)$		
$E(X_{7v})$	-2.37			
$E(X_{6c})$	2.28			
$E(X_{7c})$	2.66			
$E(L_{6v})$	-10.92	$-10.6(3)$		
$E(L_{6v})$	-6.23			
$E(L_{6v})$	-1.26			
$E(L_{4,5v})$	-1.00	$-0.9(3)$		
$E(L_{6c})$	1.53			
$E(L_{6c})$	5.42			
$E(L_{4,5c})$	5.55			
$E(\Sigma^{min})$		$-3.3(2)$		

direct energy gap:

$E_{g,dir}(\Gamma_{8v} - \Gamma_{6c})$	0.4180 eV	4.2 K	magnetotransmission	75V
	0.354(3) eV	295 K	electroreflectance	77L

For temperature dependence of energy gaps, see Fig. 2.

spin-orbit splitting energies (in eV):

$\Delta_0(\Gamma_{8v} - \Gamma_{7v})$	0.38(1)	1.5 K	magneto-electroreflectance	67P
$\Delta_1(L_{4,5v} - L_{6v})$	0.267	5 K	wavelength modulated reflectance	70Z

effective mass, electrons (in units of m_0):

m_n	0.0239		calculated from valence band parameters below	83K
	0.0231	150 K	magnetophonon resonance	81T
	0.0219(5)	250 K		

For dependence on electron concentration, see Fig. 3.

electron g-factor:

g_c	-15.0		esr	82P

For the camel's back structure at the conduction band X-minima, see [85K].

effective masses, holes (in units of m_0):

$m_{p,h}$	0.43	along [111]	calculated from valence band parameters below	83K
	0.35	along [100]		
$m_{p,l}$	0.026(2)	20 K	magnetoabsorption	67P

valence band parameters:

A	-19.7		calculated using $k \cdot p$ theory	75W
B	-16.8			
$|C|$	13.66			

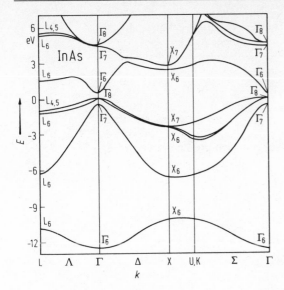

Fig. 1. InAs. Band structure obtained with a non-local pseudopotential calculation [76C].

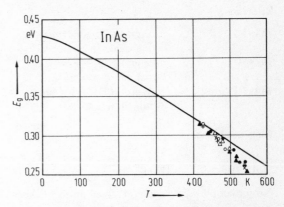

Fig. 2. InAs. Band gap vs. temperature. Solid line: optical band gap calculated from parameters used for fitting various transport data; symbols: thermal band gap of six samples obtained from conductivity and Hall coefficient [82Y].

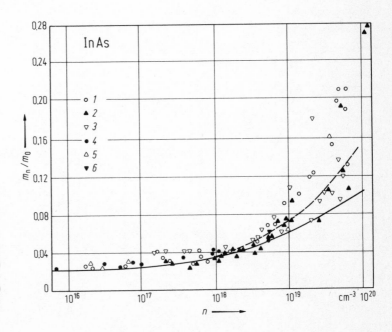

Fig. 3. InAs. Electron effective mass vs. electron concentration. Experimental data obtained by (1) Seebeck effect, (2) infrared reflectivity, (3) magnetic susceptibility, (4) Faraday effect, (5) recombination radiation, (6) cyclotron resonance [78S].

Physical property	Numerical value	Experimental conditions	Experimental method, remarks	Ref.

Lattice properties

Structure

InAs I space group $T_d^2 - F\bar{4}3m$ (zincblende structure) stable at normal pressure

InAs II space group $O_h^5 - Fm3m$ (rocksalt structure) high pressure phase 85V

InAs III space group $D_{4h}^{19} - I4_1/amd$ (β-tin structure) 85V

transition pressures (in kbar):

p_{tr}(I–II) 70(2) RT X-ray diffraction 85V
p_{tr}(II–III) 170(4)

No further phase transition observed up to 270 kbar [85V].

lattice parameter:

a 6.0583 Å 298.15 K 63O

linear thermal expansion coefficient:

α $4.52 \cdot 10^{-6} K^{-1}$ 20···250 K Figs. 4, 5 58S

density:

d $5.667 \, g \, cm^{-3}$ 300 K X-ray 69R

melting point:

T_m 1215 K 69G

phonon dispersion relations: Fig. 6.

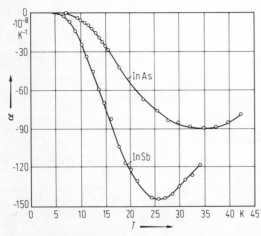

Fig. 4. InAs, InSb. Linear thermal expansion coefficient vs. temperature measured with a variable transformer dilatometer [67S].

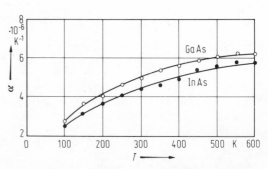

Fig. 5. GaAs, InAs. Temperature dependence of the coefficient of linear thermal expansion between 100 K and 600 K [58S].

Physical property	Numerical value	Experimental conditions	Experimental method, remarks	Ref.

phonon wavenumbers (in cm^{-1}):

$\bar{v}_{TO}(\Gamma)$	217.3	300 K	Raman scattering	80C
$\bar{v}_{LO}(\Gamma)$	238.6			
$\bar{v}_{TA}(X)$	53			
$\bar{v}_{LA}(X)$	160			
$\bar{v}_{TO}(X)$	216			
$\bar{v}_{LO}(X)$	203	100 K		
$\bar{v}_{TA}(L)$	44	300 K		
$\bar{v}_{LA}(L)$	139.5			
$\bar{v}_{TO}(L)$	216			
$\bar{v}_{LO}(L)$	203	100 K		

elastic moduli (in 10^{11} dyn cm^{-2}):

c_{11}	8.329	RT, n-type	ultrasound	63G
c_{12}	4.526			
c_{44}	3.959			

Transport properties

The transport properties are similar to those of InSb, i.e. the high electron mobility and the high mobility ratio determines the behavior of n-type and near-intrinsic samples.

intrinsic carrier concentration:
n_i can be expressed in the range $350 \cdots 900$ K by: $2.14 \cdot 10^{15}\, T^{3/2} \exp(-0.47/2\,kT)$ (kT in eV, T in K, n_i in cm^{-3}) (Fig. 7 [54F]). The extrapolated RT value then is:

n_i	$1.3 \cdot 10^{15}$ cm^{-3}	300 K		

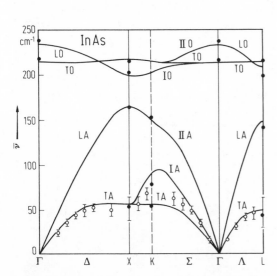

Fig. 6. InAs. Phonon dispersion relations. Solid lines: calculated, open circles: neutron data, full circles: Raman data [80C].

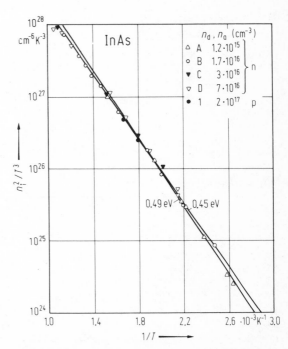

Fig. 7. InAs. Square of the intrinsic carrier concentration over T^3 vs. reciprocal temperature for different samples [54F].

Physical property	Numerical value	Experimental conditions	Experimental method, remarks	Ref.
electron and hole mobility (in cm²/Vs):				
μ_n	$0.8\cdots1\cdot10^5$	77 K	pure material ($n \approx 10^{16}$ cm^{-3})	75R
	$2\cdots3.3\cdot10^4$	300 K		
For temperature dependence, see Fig. 8.				
μ_p	$100\cdots450$	300 K	temperature dependence $\propto T^m$ with m $\approx$ 1 for low temperatures and $>$ 2 for high temperatures (Fig. 9)	54F, 63M

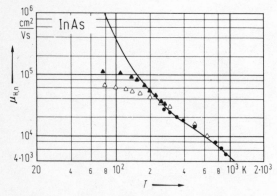

Fig. 8. InAs. Electron Hall mobility of pure material vs. temperature [75R]. Open triangles: $n=1.7\cdot10^{16}$ cm^{-3}, circles: $n=4\cdot10^{16}$ cm^{-3}, full triangles: $4\cdot10^{15}$ cm^{-3}.

thermal conductivity: Fig. 10.

Optical properties

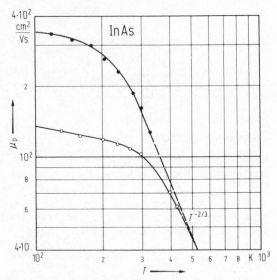

Fig. 9. InAs. Hole mobility ($8\sigma R_H/3\pi$) vs. temperature for two samples [54F].

dielectric constants:

$\varepsilon(0)$	15.15	300 K	infrared reflectance and oscillator fit	62H
$\varepsilon(\infty)$	12.25			

optical constants:
real and imaginary parts of the dielectric constant measured by spectroscopical ellipsometry; n, k, R, K calculated from these data [83A]; see also Fig. 11.

$\hbar\omega$ [eV]	ε_1	ε_2	n	k	R	K [10^3 cm^{-1}]
1.5	13.605	3.209	3.714	0.432	0.337	65.69
2.0	15.558	5.062	3.995	0.634	0.370	128.43
2.5	15.856	15.592	4.364	1.786	0.454	452.64
3.0	6.083	13.003	3.197	2.034	0.412	618.46
3.5	5.973	10.550	3.008	1.754	0.371	622.13
4.0	7.744	11.919	3.313	1.799	0.393	729.23
4.5	−1.663	22.006	3.194	3.445	0.566	1571.19
5.0	−5.923	8.752	1.524	2.871	0.583	1455.26
5.5	−3.851	6.008	1.282	2.344	0.521	1306.62
6.0	−2.403	6.005	1.434	2.112	0.448	1284.15

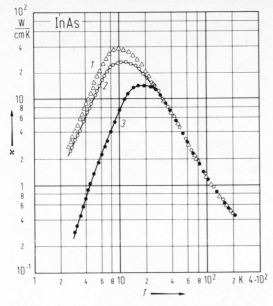

Fig. 10. InAs. Thermal conductivity below 200 K for three samples with 1: $n = 1.6 \cdot 10^{16}$ cm^{-3}, 2: $n = 2 \cdot 10^{17}$ cm^{-3}, 3: $p = 2 \cdot 10^{17}$ cm^{-3} [71T].

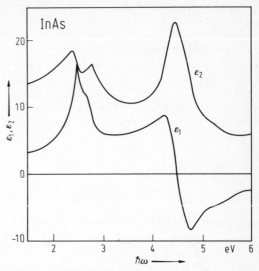

Fig. 11. InAs. Real and imaginary parts of the dielectric constant vs. photon energy [83A].

Impurities and defects

solubility of impurities in InAs

Impurity	c^{eq} [cm^{-3}]	T [°C]	Remarks	Ref.
Cd	$3.5 \cdot 10^{19}$ (max)	800	radiotracer	67A
S, Se, Te	full curves determined		microhardness and Hall effect studies	76G
Zn	$> 3 \cdot 10^{19}$ (max)	800		67B2

diffusion coefficients

Element	D_0 [cm^2 s^{-1}]	Q [eV]	Remarks	Ref.
Self-diffusion coefficients				
In	$6.0 \cdot 10^5$	4.0	radiotracer	69K
As	$3.0 \cdot 10^7$	4.45	radiotracer	69K
Impurity diffusion-coefficients				
Ag	$7.3 \cdot 10^{-4}$	0.26	radiotracer	67B1
Au	$5.8 \cdot 10^{-4}$	0.65	radiotracer	67R
Cd	$7.4 \cdot 10^{-4}$	1.15	radiotracer	67A
Cd		2.4	pn-junction	81H
Cu	$3.6 \cdot 10^{-3}$	0.52	radiotracer	67F
Ge	$3.74 \cdot 10^{-6}$	1.17	pn-junction	62S
Mg	$1.98 \cdot 10^{-6}$	1.17	pn-junction	62S
S	6.78	2.20	pn-junction	62S
Sn	$1.49 \cdot 10^{-6}$	1.17	pn-junction	62S
Se	12.6	2.20	pn-junction	62S
Te	$3.43 \cdot 10^{-5}$	1.28	pn-junction	62S
Hg	$1.45 \cdot 10^{-5}$	1.32	radiotracer	71S
Zn	$4.2 \cdot 10^{-3}$	0.96 ± 0.02	isoconcentration	67B2
Zn	complex profiles		technique not known	73C

shallow impurities

Little is known about impurities in this material. The binding energies of some acceptors and donors are determined from photoluminescence experiments. It is, however, not known whether these impurities are point defects or complexes.

acceptor binding energies

Impurity	E_b [meV]	T [K]	Remarks	Ref.
Sn	10	77	photoluminescence of implanted material	75G
Ge	14		photoluminescence	74G
Si	20			
(?)	20		photoluminescence of Sn-doped material	76Z
structure defect	35	4.2		

References for 2.14

54F   Folberth, O.G., Madelung, O., Weiss, H.: Z. Naturforsch. **9a** (1954) 954.
58S   Sirota, N.N., Pashintsev, Yu.I.: Inzh. Fiz. Zh. Akad. Nauk BSSR **1** (1958) 38.
62H   Hass, M., Henvis, B.W.: J. Phys. Chem. Solids **23** (1962) 1099.
62S   Schillmann, E.: Compound Semiconductors – Preparation of III–V Compounds, Vol. 1, Willardson, R.K., Goering, H.L. (eds.), New York: Reinhold **1962**, p. 358.
63G   Gerlich, D.: J. Appl. Phys. **34** (1963) 2915.
63M   Mikhailova, M.P., Nasledov, D.N., Slobodchikov, S.V.: Sov. Phys. Solid State (English Transl.) **5** (1964) 1685; Fiz. Tverd. Tela **5** (1963) 2317.
63O   Ozolin'sh, J.V., Averkieva, G.K., Ilvin'sh, A.F., Goryunova, N.A.: Sov. Phys. Cryst. (English Transl.) **7** (1963) 691.
67A   Arseni, K.A., Boltaks, B.I., Rembeza, S.I.: Sov. Phys. Solid State **8** (1967) 2248.
67B1   Boltaks, B.I., Rembeza, S.I., Sharma, B.L.: Sov. Phys. Solid State **1** (1967) 196.
67B2   Boltaks, B.I., Rembeza, S.I.: Sov. Phys. Solid State **8** (1967) 2117.
67P   Pidgeon, C.R., Groves, S.H., Feinleib, J.: Solid State Commun. **5** (1967) 677.
67R   Rembeza, S.I.: Sov. Phys. Solid State **1** (1967) 516.
67S   Sparks, P.W., Swenson, C.A.: Phys. Rev. **163** (1967) 779.
69G   Glazov, V.M., Chizhevskaya, S.N., Evgen'ev, S.B.: Zh. Fiz. Khim. **43** (1969) 373.
69K   Kato, H., Yokozawa, M., Kohara, R., Okabayashi, Y., Takayanagi, S.: Solid-State Electron. **12** (1969) 137.
69R   Reifenberger, B., Keck, M.J., Trivisoono, J.: J. Appl. Phys. **40** (1969) 5403.
70Z   Zucca, R.R.L., Shen, Y.R.: Phys. Rev. **155** (1970) 2668.
71S   Sharma, B.L., Purohit, R.K., Mukerjee, S.N.: J. Phys. Chem. Solids **32** (1971) 1397.
71T   Tamarin, P.V., Shalyt, S.S.: Sov. Phys. Semicond. (English Transl.) **5** (1971) 1097; Fiz. Tekh. Poluprovodn. **5** (1971) 1245.
73C   Casey, H.C.: Quoted unpublished measurements of M.G. Buehler and G.L. Pearson, in "Atomic Diffusion in Semiconductors", Shaw, D. (ed.), New York: Plenum Press **1973**, p. 351.
74G   Guseva, M.I., Zotova, N.V., Koval, A.V., Nasledov, D.N.: Sov. Phys. Semicond. (English Transl.) **8** (1974) 34; Fiz. Tekh. Poluprovodn. **8** (1974) 59.
74L   Ley, L., Pollak, R.A., McFeely, F.R., Kowalczyk, S.P., Shirley, D.A.: Phys. Rev. **B9** (1974) 600.
75G   Guseva, M.I., Zotova, N.V., Koval, A.V., Nasledov, D.N.: Sov. Phys. Semicond. (English Transl.) **8** (1975) 1323; Fiz. Tekh. Poluprovodn. **8** (1974) 2034.
75R   Rode, D.L.: in "Semiconductors and Semimetals", Vol. 10, R.K. Willardson, A.C. Beer eds., Academic Press, New York **1975**.
75V   Varfolomeev, A.V., Seisyan, R.P., Yakimova, R.N.: Sov. Phys. Semicond. (English Transl.) **9** (1975) 530; Fiz. Tekh. Poluprovodn. **9** (1975) 804.
75W1   Wiley, J.D.: in "Semiconductors and Semimetals", Vol. 10, R.K. Willardson, A.C. Beer eds., Academic Press, New York 1975.
76C   Chelikowsky, J.R., Cohen, M.L.: Phys. Rev. **B14** (1976) 556.
76G   Glazov, V.M., Akopyan, R.A., Shvedkov, E.I.: Sov. Phys.-Semicond. **10** (1976) 378.
76Z   Zotova, N.V., Karataev, V.V., Koval, A.V.: Sov. Phys. Semicond. (English Transl.) **9** (1976) 1275; Fiz. Tekh. Poluprovodn. **9** (1975) 1944.
77L   Lukeš, F.: Phys. Status Solidi (b) **84** (1977) K113.
78S   Semikolenova, N.A., Nesmelowa, I.M., Khabarov, E.N.: Sov. Phys. Semicond. (English Transl.) **12** (1978) 1139; Fiz. Tekh. Poluprovodn. **12** (1978) 1915.
80C   Carles, R., Saint-Cricq, N., Renucci, J.B., Renucci, M.A., Zwick, A.: Phys. Rev. **B22** (1980) 4804.
81H   Horikoshi, Y., Saito, H., Takanashi, Y.: Jpn. J. Appl. Phys. **20** (1981) 437.
81T   Takayama, J., Shimomae, K., Hamaguchi, C.: Jpn. J. Appl. Phys. **20** (1981) 1265.
82P   Pascher, H.: Opt. Commun. **41** (1982) 106.
82Y   Yang June Jung, Byung Ho Kim, Hyung Jae Lee, Wolley, J.C.: Phys. Rev. **26** (1982) 3151.
83A   Aspnes, D.E., Studna, A.A.: Phys. Rev. **B27** (1983) 985.

83K   Kanskaya, L.M., Kokhanovskii, S.I., Seisyan, R.P., Efros, Al.L., Yukish, V.A.: Sov. Phys. Semicond. (English Transl.)
       17 (1983) 449; Fiz. Tekh. Poluprovodn. **17** (1983) 718.
83W   Williams, G.P., Cerrina, F., Anderson, J., Lapeyre, G.J., Smith, R.J., Hermanson, J., Knapp, J.A.: Physica **117B &**
       118B (1983) 350.
85V   Vohra, Y.K., Weir, S.T., Ruoff, A.L.: Phys. Rev. **B31** (1985) 7344.

Physical property	Numerical value	Experimental conditions	Experimental method, remarks	Ref.

2.15 Indium antimonide (InSb)

Electronic properties

band structure: Fig. 1 (Brillouin zone: see Fig. 2 of section 1.1)

InSb is a direct semiconductor. The minimum of the conduction band (Γ_6) is situated in the center of the Brillouin zone. Near the minimum, $E(k)$ is isotropic but non-parabolic. Thus the effective mass of the electrons is scalar and depends strongly on the electron concentration. Higher band minima (about 0.63 eV above the lowest minimum) seem to be established by transport measurements in heavily doped n-InSb [75F]. The valence band shows the structure common to all zincblende semiconductors i.e. two subbands degenerate at Γ_8 and one spin-split band (Γ_7). A small crystal field splitting of the heavy hole band is negligible for most phenomena.

energies of symmetry points of the band structure (relative to the top of the valence band) (in eV):

$E(\Gamma_{6v})$	-11.71	-11.7 (d)	symmetry symbols in double
$E(\Gamma_{7v})$	-0.82	-0.850 (a)	group notation
$E(\Gamma_{8v})$	0.00		first row: theoretical data
$E(\Gamma_{6c})$	0.25	0.235 (b)	of [76C, 85C], see Fig. 1
$E(\Gamma_{7c})$	3.16	3.141 (a)	second row: experimental
$E(\Gamma_{8c})$	3.59	3.533 (a, c)	data deduced from (a):[85L1],
$E(X_{6v})$	-9.20	-9.5 (d)	(b):[83L], (c):[81M]),
$E(X_{6v})$	-6.43	-6.4 (d)	(d):[74L]
$E(X_{6v})$	-2.45	-2.4 (d)	
$E(X_{7v})$	-2.24		
$E(X_{6c})$	1.71	1.79 (a, d)	
$E(X_{7c})$	1.83		
$E(L_{6v})$	-9.95	-10.5 (d)	
$E(L_{6v})$	-5.92		

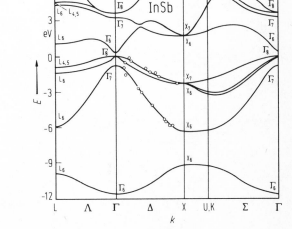

Fig. 1. InSb. Band structure obtained with a non-local pseudopotential calculation [76C], corrected in [84C] (Fig. from [84C]). Experimental data from angular resolved photoemission from a InSb (001) surface [83H] have been included (circles).

Physical property	Numerical value	Experimental conditions	Experimental method, remarks	Ref.
$E(L_{6v})$	-1.44	-1.4 (d)		
$E(L_{4,5v})$	-0.96	-0.9 (a, d)		
$E(L_{6c})$	1.03			
$E(L_{6c})$	4.30	4.32 (a, d)		
$E(L_{4,5c})$	4.53	4.47 (a, d)		

energy gaps (in eV):

$E(1S)$	$0.2363(2)$	$2\,K, n_{77K} = 6\cdot10^{13}\,cm^{-3}$	exciton ground state, from luminescence and absorption	79K2
$E_{g,dir}$	$0.2368(2)$		calculated from $E(1S)$	79K2
	0.2352	$1.8\,K$	resonant two-photon photo-Hall effect,	85L2,
	0.230	$77\,K$	two-photon magnetoabsorption	82G

temperature dependence of energy gap:

$E_g(T)$	$E_g(0) - aT^2/(b+T)$ with $a = 0.6\,meV\,K^{-1}, b = 500\,K$, Fig. 2			85L2

For camel's back structure of the conduction band X-minima, see [85K].

spin-orbit splitting energies (in eV):

$\Delta_0(\Gamma_{8v} - \Gamma_{7v})$	0.850	$100\,K$	ellipsometry	85L1
$\Delta_1(\Lambda_{4,5v} - \Lambda_{6v})$	$0.498(4)$			
$\Delta_0'(\Gamma_{7c} - \Gamma_{8c})$	$0.392(12)$		ellipsometry	81M

critical point energies (in eV):

$E_1(\Lambda_{4,5v} - \Lambda_{6c})$	$1.968(1)$	$100\,K$	ellipsometry	85L1
	$1.872(2)$	RT		81M
$E_0'(\Gamma_{8v} - \Gamma_{7c})$	$3.141(12)$			85L1
$E_2(X_{6,7v} - X_{6c})$	$4.186(2)$			
$E_1'(\Lambda_{4,5v} - L_{6c})$	$5.22(3)$			

effective masses, electrons (in units of m_0):

$m_n(\Gamma_{6c})$	$0.01359(3)$	$4.2\,K, n = 4.6 \cdot10^{13}\,cm^{-3}$	Faraday effect	83Z

Dependence of electron effective mass on carrier concentration: see Fig. 3; energy dependence caused by the non-parabolicity of the conduction band: see Fig. 4.

$m_n(L_{6c})$	$0.09\,m_0$		electroreflection	79Z

electron g-factor:

g_c	-50.6		intra conduction band magnetoabsorption	83G

effective masses, holes (in units of m_0):

$m_{p,h}$	$0.45(3)$	$T = 4\cdots77\,K$, $\parallel[111]$	cyclotron resonance	63B2
	$0.42(3)$	$\parallel[110]$		
	$0.34(3)$	$\parallel[100]$		
$m_{p,1}$	$0.0158(5)$	$77\,K$, p-type	magnetoplasma resonance (decreases to 0.0147 at 150 K)	80S

Physical property	Numerical value	Experimental conditions	Experimental method, remarks	Ref.
valence band parameters:				
A	-35		calculated using $k \cdot p$ theory	75W
B	-31.4			
$|C|$	20.92			

Lattice properties

Structure

InSb I	space group T_d^2–F$\overline{4}$3m (zincblende structure)	stable at normal pressure	

Data on high-pressure phases are conflicting [78Y].

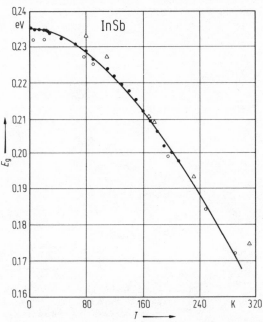

Fig. 2. InSb. Energy gap vs. temperature below RT measured by resonant two-photon photo-Hall effect (full circles); open circles and triangles: earlier literature data for comparison. Solid curve: fit by Varshni's formula as given in the tables [85L2].

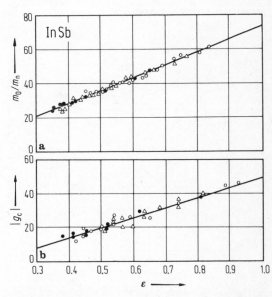

Fig. 4. InSb. Dependence of electron effective mass and g-factor on energy in the conduction band $\Delta E = E - E_c$. (a) reciprocal mass and (b) g-factor vs. $\varepsilon = E_g/(E_g + 2\Delta E)$ [83K].

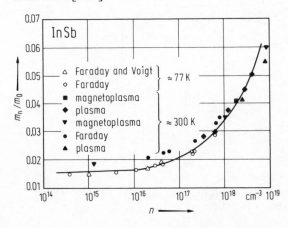

Fig. 3. InSb. Electron effective mass vs. carrier concentration. Comparison of results of several experimental measurements. Solid line: Kane's theory [77S].

Physical property	Numerical value	Experimental conditions	Experimental method, remarks	Ref.

lattice parameter:

a	6.47937 Å	298.15 K	X-ray	65S

Temperature dependence in the range $10 \cdots 60\,^\circ$C and coefficient of linear thermal expansion, see Figs. 5 and 6.

density:

d	5.7747 (4) g cm^{-3}		X-ray	65S

melting point:

T_m	800 (3) K			73H

phonon dispersion relations: Fig. 7.

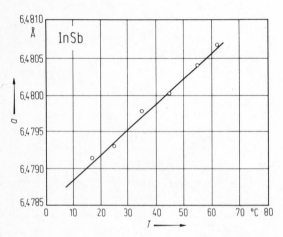

Fig. 5. InSb. Lattice parameter vs. temperature [65S].

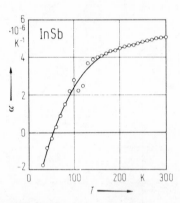

Fig. 6. InSb. Linear thermal expansion coefficient vs. temperature measured with an interferometer [58G]. High temperature range.

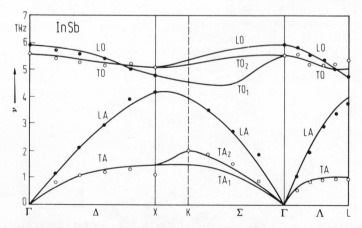

Fig. 7. InSb. Phonon dispersion relations calculated with a six parameter dynamic model [85R]. Experimental points from [71P].

Physical property	Numerical value	Experimental conditions	Experimental method, remarks	Ref.
phonon frequencies (in THz):				
$\nu_{LO}(\Gamma_{15})$	5.90(25)	300 K, $n = 8 \cdot 10^{13}\,cm^{-3}$	coherent inelastic neutron scattering	71P
$\nu_{TO}(\Gamma_{15})$	5.54(5)			
$\nu_{TA}(X_5)$	1.12(5)			
$\nu_{LA}(X_3)$	4.30(10)			
$\nu_{LO}(X_1)$	4.75(20)			
$\nu_{TO}(X_5)$	5.38(17)			
$\nu_{TA}(L_3)$	0.98(5)			
$\nu_{LA}(L_1)$	3.81(6)			
$\nu_{LO}(L_1)$	4.82(10)			
$\nu_{TO}(L_3)$	5.31(6)			
$\nu_{TO}(\Gamma)$	5.49	0 K, extrapol.	zone center phonons, Raman spectroscopy	84L
	5.39	RT		
$\nu_{LO}(\Gamma)$	5.83	0 K, extrapol.		
	5.72	RT		

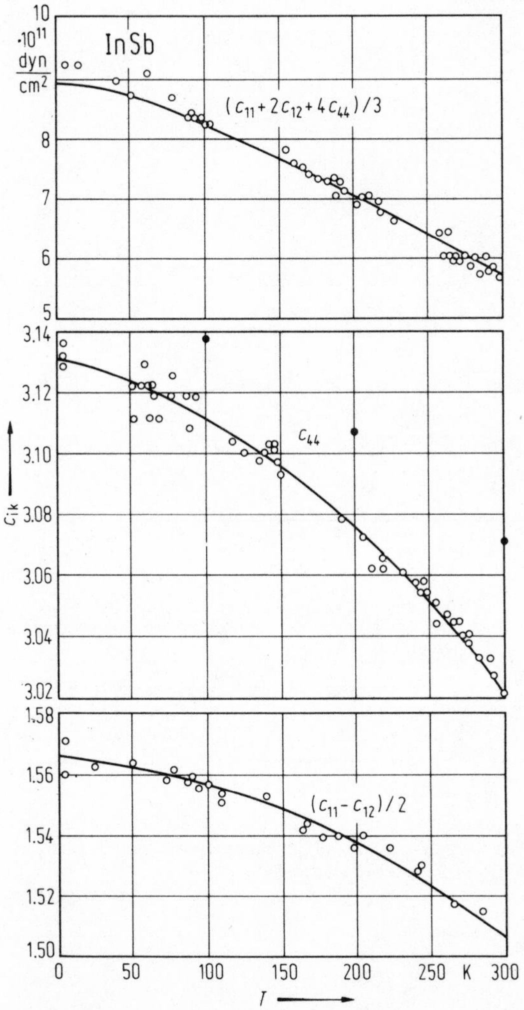

Fig. 8. InSb. Second order elastic moduli vs. temperature [59S].

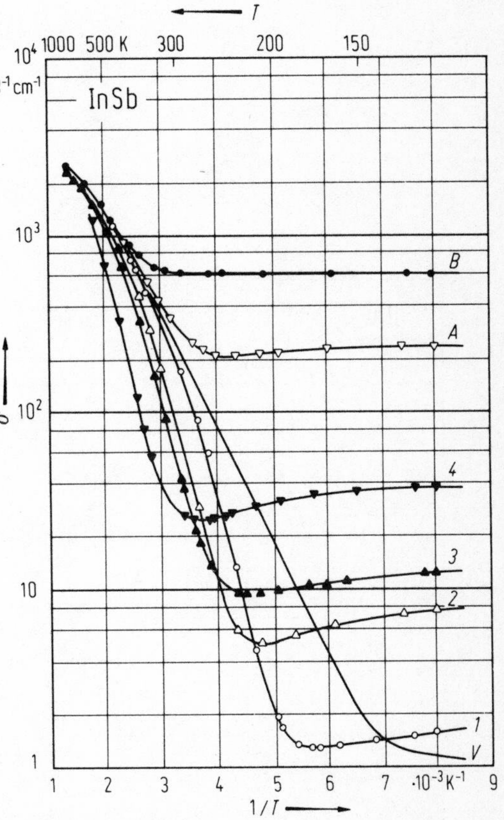

Fig. 9. InSb. Conductivity vs. reciprocal temperature. Curve V: $n \approx 10^{13}\,cm^{-3}$ [59V], A: $n = 1.3 \cdot 10^{16}\,cm^{-3}$, B: $n = 1 \cdot 10^{16}\,cm^{-3}$, 1: $p = 4 \cdot 10^{15}\,cm^{-3}$, 2: $p = 2.2 \cdot 10^{16}\,cm^{-3}$, 3: $p = 6 \cdot 10^{16}\,cm^{-3}$, 4: $p = 2 \cdot 10^{17}\,cm^{-3}$ [54M].

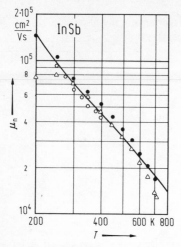

Fig. 10. InSb. Electron mobility vs. temperature. Experimental data are Hall mobilities from various sources. Solid line: calculated drift mobility [71R].

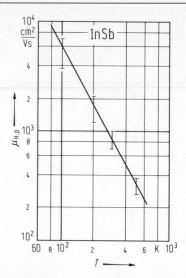

Fig. 11. InSb. Hole Hall mobility vs. temperature. The vertical bars indicate the ranges for the experimental values. The solid line has the slope −1.8 [75W].

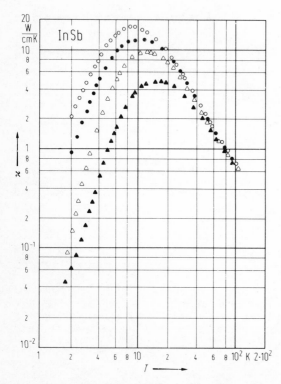

Fig. 12. InSb. Thermal conductivity vs. temperature of several p-type samples [71K].

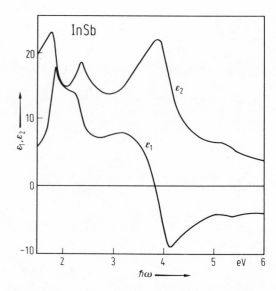

Fig. 13. InSb. Real and imaginary parts of the dielectric constant vs. photon energy [83A].

Physical property	Numerical value	Experimental conditions	Experimental method, remarks	Ref.

second order elastic moduli (in $10^{11}\,\mathrm{dyn\,cm^{-2}}$):

c_{11}	6.918	0 K, extrapol.	ultrasound	59S
	6.669	300 K	for temperature dependence,	
c_{12}	3.788	0 K, extrapol.	see Fig. 8.	
	3.645	300 K		
c_{44}	3.132	0 K, extrapol.		
	3.020	300 K		

Transport properties

The transport properties in n-type material are mainly determined by the extremely high mobility of the electrons in the lowest conduction band minimum. Pure material with intrinsic conduction down to 200 K is available. At low temperatures and in doped material ionized-impurity scattering limits the electron mobility. In pure material below RT polar optical mode scattering is dominant. Above 400 K electron-hole scattering contributes significantly. Above 700 K multivalley conduction seems to occur.

electrical conductivity: Fig. 9.

intrinsic carrier concentration:

n_i	$2.04\,(55)\cdot 10^{16}\,\mathrm{cm^{-3}}$	300 K	Hall effect	70C

The temperature dependence can be described in the range $150\cdots 300\,\mathrm{K}$ by the relation:

$$n_i = 5.76\cdot 10^{14}\,T^{3/2}\exp(-0.26/2\,kT)\,\mathrm{cm^{-3}}\;(T\text{ in K}, kT\text{ in eV})[70C]$$

electron mobility (in $\mathrm{cm^2/Vs}$):

$\mu_{H,n}$	$7.7\cdot 10^4\cdot (T/300\,\mathrm{K})^{-1.66}$		see also Fig. 10	64M
	$6\cdot 10^5$	$n_d = 5\cdot 10^{14}\,\mathrm{cm^{-3}}$, 77 K		85W
$\mu_{dr,n}$	$5.25\cdot 10^5$			

hole mobility (in $\mathrm{cm^2/Vs}$):

$\mu_{p,h}$	$850\,(T/300\,\mathrm{K})^{-b}$		b is $1.8\cdots 2.1$ for the range $60\cdots 500\,\mathrm{K}$, see Fig. 11.	75W
$\mu_{p,1}$	$3\cdot 10^4$	295 K	from analysis of transport data	64S

thermal conductivity: Fig. 12.

Optical properties

dielectric constants:

$\varepsilon(0)$	16.8 (2)		gyroscopic sphere resonance	80D
	$17.3\cdots 18.0$		range of results obtained by infrared reflectivity (from compilation of data in [80D])	
$\varepsilon(\infty)$	15.68		infrared reflectance and oscillator fit	62H

optical constants

real and imaginary parts of the dielectric constant measured by spectroscopical ellipsometry; n, k, R, K calculated from these data [83A]; see also Fig. 13.

$\hbar\omega$[eV]	ε_1	ε_2	n	k	R	$K\,[10^3\,cm^{-1}]$
1.5	19.105	5.683	4.418	0.643	0.406	97.79
2.0	14.448	14.875	4.194	1.773	0.443	359.46
2.5	7.811	15.856	3.570	2.221	0.447	562.77
3.0	7.354	13.421	3.366	1.994	0.416	606.27
3.5	5.995	17.673	3.511	2.517	0.474	892.82
4.0	−6.722	19.443	2.632	3.694	0.608	1497.79
4.5	−6.297	8.351	1.443	2.894	0.598	1320.24
5.0	−4.250	6.378	1.307	2.441	0.537	1237.01
5.5	−4.325	4.931	1.057	2.333	0.563	1300.55
6.0	−3.835	3.681	0.861	2.139	0.572	1300.85

Impurities and defects

diffusion coefficients

Element	$D_0[cm^2\,s^{-1}]$	Q[eV]	T[K]	Remarks	Ref.
Self-diffusion coefficients in InSb.					
In	$5.0\cdot10^{-2}$	1.82	480···520	radiotracer (residual activity)	57E
In	$1.8\cdot10^{13}$	4.3	475···517	radiotracer	68K
Sb	$5.0\cdot10^{-2}$	1.94	480···520	radiotracer (residual activity)	57E
Sb	$3.1\cdot10^{13}$	4.3	475···517	radiotracer	68K
Impurity diffusion coefficients					
Ag	$\approx1.0\cdot10^{-7}$	≈0.25		radiotracer	62W
Au	$7.0\cdot10^{-4}$	0.32		radiotracer	64B
Co	$\approx1.0\cdot10^{-7}$	0.25		radiotracer	62W
Cd	$1.0\cdot10^{-5}$	1.1		radiotracer	63B1
Cu	$9.0\cdot10^{-4}$	1.08		radiotracer	63S
Fe	$\approx1.0\cdot10^{-7}$	≈0.25		radiotracer	62W
F	22	1.0		ion analyzer	79B
Hg	$4.0\cdot10^{-6}$	1.17		radiotracer	64G
Li	$7.0\cdot10^{-4}$	0.28		electrical profile	66T
Se	1.6	1.87		CV profiling	69R
Sn	$5.5\cdot10^{-8}$	0.75		radiotracer	61S
Te	$1.7\cdot10^{-7}$	0.57		radiotracer	57B
Zn	non-ideal profiles			radiotracer	73C

donors

Shallow, effective-mass like donors are always present in concentrations of at least $10^{13}\,cm^{-3}$. Their influence on the semiconductor properties is strong due to their small binding energy of $\approx0.7\,meV$. No detailed investigations exist.

acceptors

acceptor binding energies

Impurity	E_b [meV]	T [K]	Remarks	Ref.
Theory	8.5		$1S_{3/2}$, acceptor effective mass calculation	73L
Cd	9.86	4.5	Fourier transform spectroscopy	73K
Zn	9.86			
	9.1	1.5	infrared absorption	72M

Impurity	E_b [meV]	T [K]	Remarks	Ref.
Ge	9.25			
Cr	70	14···140	Hall effect, unknown charge; state acts as acceptor	80P
Mn	9.5	4.2···77	Hall effect	71D
Fe	13	10···200	Hall effect	73V1
Co	8	4.2···200	Hall effect	73V2
Cu^0	28	4.2···300	Hall effect	79K1
Cu^-	56			
Ag^0	29.8	4.5	Fourier transform spectroscopy, weakly bound hole	73K
Ag^-	56	4.2···300	Hall effect	79K1

References for 2.15

54M Madelung, O., Weiss, H.: Z. Naturforsch. **9a** (1954) 527.
57B Boltaks, B.I., Kulikov, G.S.: Sov. Phys. Tech. Phys. **2** (1957) 67.
57E Eisen, F.H., Birchenall, C.E.: Acta Metall. **5** (1957) 265.
58G Gibbons, D.F.: Phys. Rev. **112** (1958) 779.
59P Putley, E.H.: Proc. Phys. Soc. **73** (1959) 128.
59S Slutsky, L.J., Garland, C.W.: Phys. Rev. **113** (1959) 167.
59V Volokobinskaya, N.I., Galavanov, V.V., Nasledov, D.N.: Sov. Phys. Solid State (English Transl.) **1** (1959) 687; Fiz. Tverd. Tela **1** (1959) 756.
61S Sze, S.M., Wei, L.Y.: Phys. Rev. **124** (1961) 84.
62H Hass, M., Henvis, B.W.: J. Phys. Chem. Solids **23** (1962) 1099.
62W Watt, L.A.K., Chen, W.S.: Bull. Am. Phys. Soc. **7** (1962) 89.
63B1 Boltaks, B.I., Sokolov, V.I.: Sov. Phys. Solid State **5** (1963) 785.
63B2 Bagguley, D.M.S., Robinson, M.L.A., Stradling, R.A.: Phys. Lett. **6** (1963) 143.
63S Stocker, H.J.: Phys. Rev. **130** (1963) 2160.
64B Boltaks, B.I., Sokolov, V.I.: Sov. Phys. solid State **6** (1964) 600.
64G Gusev, I.A., Murin, A.N.: Sov. Phys. Solid State **6** (1964) 1229.
64M Madelung, O.: Physics of III-V Compounds, J. Wiley & Sons, New York **1964**.
64S Schönwald, H., Z. Naturforsch. **19a** (1964) 1276.
65S Straumanis, M.E., Kim, C.D.: J. Appl. Phys. **36** (1965) 3822.
66T Takobutake, T., Ikari, H., Uyeda, Y.: Jpn. J. Appl. Phys. **5** (1966) 839.
68K Kendall, D.L.: Semiconductors and Semimetals, Vol. 4, Willardson, R.K., Beer, A.C. (eds.), New York, London: Academic Press **1968**, p. 163.
69R Rekalova, G.I., Shokov, A.A., Gavrushko, V.V.: Sov. Phys. Semicond. **2** (1969) 1452.
70C Cunningham, R.W., Gruber, J.B.: J. Appl. Phys. **41** (1970) 1804.
71D Dashevskii, M.Ya., Ivleva, V.S., Korl, L.Ya., Kurilenko, I.N., Litvak-Gorskaya, L.B., Mitrofanova, R.S., Fridlyand, E.Yu.: Sov. Phys. Semicond. (English Transl.) **5** (1971) 757; Fiz. Tekh. Poluprovodn. **5** (1971) 858.
71K Kosarev, V.V., Tamarin, P.V., Shalyt, S.S.: Phys. Status Solidi (b) **44** (1971) 525.
71P Price, D.L., Rowe, J.M., Nicklow, R.M.: Phys. Rev. **B3** (1971) 1268.
71R Rode, D.L.: Phys. Rev. **B3** (1971) 3287.
72M Murzin, V.N., Demishina, A.I., Umarov, L.M.: Sov. Phys. Semicond. (English Transl.) **6** (1972) 419; Fiz. Tekh. Poluprovodn. **6** (1972) 488.
73C Unpublished results by D.L. Kendall, quoted by Casey, H.C.: Atomic Diffusion in Semiconductors, Shaw, D. (ed.), New York: Plenum Press **1973**, p. 351.
73H Hultgren, R., Desai, P.D., Hawkins, D.T., Gleiser, M., Kelly, K.K., Wagman, D.D.: Selected Values of the Thermodynamic Properties of Binary Alloys, American Society of Metals, Metal Park, Ohio **1973**.
73K Kaplan, R.: Solid State Commun. **12** (1973) 191.
73L Lipari, N.O.: Ref. 6 in [73K].
73V1 Vinogradova, K.I., Ivleva, V.S., Il'menkov, G.V., Nasledov, D.N., Smetannikova, Yu.S., Tashkhodshaev, T.K.: Sov. Phys. Semicond. (English Transl.) **6** (1973) 1595; Fiz. Tekh. Poluprovodn. **6** (1973) 1845.
73V2 Vinogradova, K.I., Nasledov, D.N., Smetannikova, Yu.S., Tashkhodshaev, T.K.: Sov. Phys. Solid State (English Transl.) **15** (1973) 212; Fiz. Tverd. Tela **15** (1973) 295.
74L Ley, L., Pollak, R.A., McFeely, F.R., Kowalczyk, S.P., Shirley, D.A.: Phys. Rev. **B9** (1974) 600.
75F Filipchenko, A.S., Nasledov, D.N.: Phys. Status Solidi (a) **27** (1975) 11.
75W Wiley, J.D.: in "Semiconductors and Semimetals", Vol. 10, R.K. Willardson. A.C. Beer eds., Academic Press, New York and London, **1975**.
76C Chelikowski, J.R., Cohen, M.L.: Phys. Rev. **B14** (1976) 556.

77S   Stillman, G.E., Wolfe, C.M., Dimmock, J.O.: in "Semiconductors and Semimetals", Vol. 12, R.K. Willardson, A.C. Beer eds., Academic Press, New York and London, 1977.
78Y   Yu, S.C., Spain, I.L., Skelton, E.F.: J. Appl. Phys. 49 (1978) 4741.
79B   Blaut-Blachev, A.N., Ivleva, V.S., Selyanina, V.I.: Sov. Phys. Semicond. 13 (1979) 1342.
79K1   Kurilenko, I.N., Litvak-Gorskaya, L.B., Lugovaya, G.E.: Sov. Phys. Semicond. (English Transl.) 13 (1979) 906; Fiz. Tekh. Poluprovodn. 13 (1979) 1556.
79K2   Kaskaya, L.M., Kokhanovskii, S.I., Seisyan, R.P.,: Sov. Phys. Semicond. (English Transl.) 13 (1979) 234; Fiz. Tekh. Poluprovodn. 13 (1979) 2424.
79Z   Zvonkov, B.N., Salashchenko, N.N., Filatov, O.N.: Sov. Phys. Solid State (English Transl.) 21 (1979) 777; Fiz. Tverd. Tela 21 (1979) 1344.
80D   Dixon, J.R., Furdyna, J.K.: Solid State Commun. 35 (1980) 195.
80P   Propov, V.V., Kosarev, V.V.: Phys. Status Solidi (a) 58 (1980) 231.
80S   Seiler, D.G., Goodwin, M.W., Miller, A.: Phys. Rev. Lett. 44 (1980) 807.
81M   Mattausch, H.J., Aspnes, D.E.: Phys. Rev. B23 (1981) 1896.
82G   Goodwin, M.W., Seiler, D.G., Weiler, M.H.: Phys. Rev. B25 (1982) 6300.
83A   Aspnes, D.E., Studna, A.A.: Phys. Rev. B27 (1983) 985.
83G   Goodwin, M.W., Seiler, D.G.: Phys. Rev. B27 (1982) 3451.
83H   Höchst, H., Hernández-Calderón, I.: Surf. Sci. 126 (1983) 25.
83K   Kanskaya, L.M., Kokhanovskii, S.I., Seisyan, R.P., Efros, Al.L.: Phys. Status Solidi (b) 118 (1983) 447.
83L   Littler, C.L., Seiler, D.G., Kaplan, R., Wagner, R.J.: Phys. Rev. B27 (1983) 7473.
83Z   Zengin, D.M.: J. Phys. D16 (1983) 653.
84C   Chelikowski, J.R., Cohen, M.L.: Phys. Rev. B30 (1984) 4828.
84L   Liarokapis, W., Anastassakis, E.: Phys. Rev. B30 (1984) 2270.
85K   Kopylov, A.A.: Solid State Commun. 56 (1985) 1.
85L1   Logothetidis, S., Vina, L., Cardona, M.: Phys. Rev. B31 (1985) 947.
85L2   Littler, C.L., Seiler, D.G.: Appl. Phys. Lett. 46 (1985) 986.
85R   Ram, R.K., Kushwara, S.S.: J. Phys. Soc. Jpn. 54 (1985) 617.
85W   Warmenbol, P. Peeters, F.M., Devreese, J.T., Algebra, G.E., van Welzenis, R.G.: Phys. Rev. B31 (1985) 5285.

Physical property	Numerical value	Experimental conditions	Experimental method, remarks	Ref.

2.16 Ternary and quaternary alloys between III–V compounds

Solid solutions between III–V compounds have become increasingly important for microelectronical applications, giving the possibility to choose the energy range e.g. for optoelectronic or laser applications more appropriate than with the binary compounds. The recent progress has mostly been achieved for epitaxial layers on substrates of III–V binary compounds. Fig. 1 shows the lattice parameter vs. the energy gap of various III–V compounds and their ternary and quaternary alloys. The alloys appropriate for lattice matching on GaSb, InP and GaAs and the energy gaps available with such alloys can easily be found. We discuss such alloys in the following sections.

2.16.1 Ternary alloys of the type III_x–III_{1-x}–V

Aluminum gallium nitride ($Al_xGa_{1-x}N$)
Both constituents of this pseudo-binary system are direct gap semiconductors, crystallizing in the same wurtzite structure. They are completely miscible. Most investigations have been made on thin films on sapphire or silicon.

Aluminum gallium phosphide ($Al_xGa_{1-x}P$)
Both components are indirect semiconductors with zincblende structure.

energy gap (in eV):

$E_{g,ind}$	$2.28 + 0.16\,x$	81B3

Aluminum gallium arsenide ($Al_xGa_{1-x}As$)

Electronic properties

The conduction band structure is characterized by a crossover from the sequence $\Gamma - L - X$ in GaAs to $X - L - \Gamma$ in AlAs. Thus $Al_xGa_{1-x}As$ is a direct gap semiconductor below a crossover concentration x_c and an indirect gap (X-conduction band) semiconductor at high x.

Physical property	Numerical value	Experimental conditions	Experimental method, remarks	Ref.

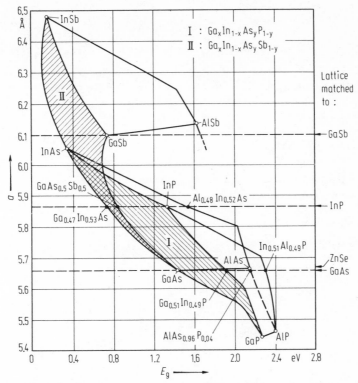

Fig. 1. Lattice parameter vs. energy gap (RT values) for various III-V compounds and their alloys.

energy gaps (energies of the Γ_{6c}, X_{6c} and L_{6c} conduction band minima relative to the top of the valence band Γ_{8v}, in eV):

$E_g(\Gamma)$	$1.420 + 1.087\,x$ $+ 0.438\,x^2$		81S
$E_g(X)$	$1.905 + 0.10\,x$ $+ 0.16\,x^2$		
$E_g(L)$	$1.705 + 0.695\,x$		
$x_c(L-X)$	0.35	crossover of X and L bands	
$x_c(\Gamma-X)$	0.43	crossover of Γ and X bands	
$x_c(\Gamma-L)$	0.47	crossover of Γ and L bands	

effective masses

According to [85A] a linear compositional dependence seems to be a good approximation.

Lattice properties

The lattice parameters of GaAs and AlAs are very similar. Thus $Al_xGa_{1-x}As$ layers on semiinsulating GaAs can be grown by MBE, LPE or VPE.

lattice parameter (in Å):

a	$5.6533 + 0.0078\,x$	85A

Physical property	Numerical value	Experimental conditions	Experimental method, remarks	Ref.
density (in $\mathrm{g\,cm^{-3}}$):				
ϱ	$5.36 - 1.6\,x$			85A
melting point (in K):				
T_m	$1511 - 58\,x + 560\,x^2$		solidus curve	85A
	$1511 + 1082\,x - 580\,x^2$		liquidus curve	
phonon wavenumbers (in $\mathrm{cm^{-1}}$):				
$\bar{\nu}_{LO}$(GaAs-type)	$292.2 - 52.8\,x + 14.4\,x^2$		best fit to experimental data of [79K]	
$\bar{\nu}_{LO}$(AlAs-type)	$359.7 + 70.8\,x - 26.8\,x^2$			
$\bar{\nu}_{TO}$(GaAs-type)	$268.3 - 5.2\,x - 9.3\,x^2$			
$\bar{\nu}_{TO}$(AlAs-type)	$359.7 + 4.4\,x - 2.4\,x^2$			
second order elastic moduli (in $10^{11}\,\mathrm{dyn\,cm^{-2}}$):				
c_{11}	$11.88 + 0.14\,x$			85A
c_{12}	$5.38 + 0.32\,x$			
c_{44}	$5.94 - 0.05\,x$			

Transport properties

The transport properties in n-type samples with x around x_c are determined by the occurrence of electrons in all three subbands of the conduction band.

thermal resistivity (in $\mathrm{K\,cm\,W^{-1}}$):				
κ^{-1}	$2.27 + 20.83\,x - 30\,x^2$			85A

Aluminum gallium antimonide ($\mathrm{Al_x Ga_{1-x}Sb}$)

The band structure of this compound is characterized by a double change of symmetry of the lowest conduction band minima from Γ to L and from L to X with increasing x [83A].

energy gaps (in eV):

$E_0(\Gamma_{8v} - \Gamma_{6c})$	$0.813 + 1.097\,x$ $+ 0.40\,x^2$	4.2 K	photoluminescence	83B2
	$0.81 + 1.09\,x + 0.48\,x^2$	27 K	modulation spectroscopy	83A
	$0.73 + 1.10\,x + 0.47\,x^2$	300 K		
$E_x(\Gamma_{8v} - X_{6c})$	$1.11 + 0.577\,x$	4.2 K		83B2
	$1.12 + 0.56\,x$	27 K		83A
	$1.05 + 0.56\,x$	300 K		
$E_L(\Gamma_{8v} - L_{6c})$	$0.893 + 0.867\,x$ $+ 0.21\,x^2$	4.2 K		83B2

Aluminum indium phosphide ($\mathrm{Al_x In_{1-x}P}$)

According to [70L1, 70O] linear x-dependence of the energy gaps (relative to the top of the valence band at Γ) (in eV):

$E_g(\Gamma)$	$1.34 + 2.23\,x$
$E_g(X)$	$2.24 + 0.18\,x$
x_c	0.44

Physical property	Numerical value	Experimental conditions	Experimental method, remarks	Ref.

Aluminum indium arsenide ($Al_xIn_{1-x}As$)

As in $Al_xIn_{1-x}P$ the band gap of $Al_xIn_{1-x}As$ changes from a direct gap in InAs to an indirect gap (at X) in AlAs [70L2].

energy gaps (in eV, relative to the top of the valence band):

$E_g(\Gamma)$	$0.37 + 1.91\,x + 0.74\,x^2$	300 K	extrapolation from data below for $x \approx 0.48$ and data for InAs and AlAs	84W
$E_g(X)$	$1.8 + 0.4\,x$			70L2

crossover concentration:

$x_c(\Gamma - X)$	0.68	300 K	$E_g(x_c) = 2.05\,eV$	70L2

$Al_{0.48}In_{0.52}As$
For $x = 0.48$ $Al_xIn_{1-x}P$ is lattice matched to InP.

$E_g(x = 0.477)$	1.450	300 K	(see Fig. 1)	84D,
	1.508	4 K		84W

Aluminum indium antimonide ($Al_xIn_{1-x}Sb$)

The band gap changes from direct to indirect as in $Al_xIn_{1-x}P$ and $Al_xIn_{1-x}As$.

Gallium indium phosphide ($Ga_xIn_{1-x}P$)

This series of alloys comprises large gap semiconductors with a direct gap for x up to about 0.7 and an indirect gap (at X) at higher values of x. An intermediate range with the L-minima as lowest conduction bands seems to exist. Of most interest are alloys with x around 0.55 lattice matched to GaAs (see Fig. 1).

energy gaps (in eV, relative to the top of the valence band):
   From a compilation of various literature data in [83B3] the bowing parameters (in eV) and crossover concentrations lie in the following ranges:

$c(\Gamma)$	$0.44 \cdots 0.88$
$c(X)$	$0.15 \cdots 0.20$
$c(L)$	$0.34 \cdots 0.86$
$x_c(\Gamma - L)$	$0.63 \cdots 0.68$
$x_c(L - X)$	$0.74 \cdots 0.77$

Data for lattice matched layers:
Values for the compositional parameter at lattice match lie in the range 0.51 to 0.56.

$E_g(\Gamma)$	1.903	x = 0.56, RT VPE-layer	photoluminescence	85Z
	$1.295 + 1.151\,x$	$0.50 \leqq x \leqq 0.53$ MBE-layer	photoluminescence	81R
	$1.469 + 0.511\,x$ $+ 0.6043\,x^2$	$0.49 \leqq x \leqq 0.55$ VPE-layer		78O
	1.902	x = 0.515, OMVPE layer	photoluminescence	85K

Gallium indium arsenide ($Ga_xIn_{1-x}As$)

Both components, InAs and GaAs, are direct gap semiconductors. Being a member of the quaternary system $Ga_xIn_{1-x}As_yP_{1-y}(y = 1)$, $Ga_xIn_{1-x}As$ is an important material for many applications in microelectronics. $Ga_{0.47}In_{0.53}As$ has the same lattice parameter as InP (see Fig. 1).

Physical property	Numerical value	Experimental conditions	Experimental method, remarks	Ref.
energy gap (in eV):				
$E_g(\Gamma)$	$0.422 + 0.7\,x + 0.4\,x^2$	2 K	photoluminescence	83B1
	$0.324 + 0.7\,x + 0.4\,x^2$	300 K		

$Ga_{0.47}In_{0.53}As$

As a member of the InP lattice matched quaternary $Ga_xIn_{1-x}As_yP_{1-y}$ alloys, $Ga_{0.47}In_{0.53}As$ has been investigated very intensively.

energy gap (in eV):				
$E_g(\Gamma)$	0.813	2 K	photoluminescence	83G
	0.75	300 K		
temperature dependence of energy gap:				
$E_g(T)$	$0.814 - 4.906\cdot10^{-4}\,T^2/(T + 301)$		photoluminescence, T in K	84Y
energy difference between Γ and L conduction bands:				
$E(L_{6c}) - E(\Gamma_{6c})$	0.55(5) eV	300 K	photoemission	82C
effective mass, electrons:				
m_n	$0.041\,m_0$	0 K, extrapol.	cyclotron resonance	85S
effective masses, holes (in units of m_0):				
$m_{p,h}$	0.465	‖ [001]		80A
	0.56	‖ [110]		
	0.60	‖ [111]		
$m_{p,l}$	0.0503			
phonon wavenumbers (in cm^{-1}):				
$\bar{v}$	226	TO	Raman active modes	83P
	270.5	LO		
	235	LO, InAs-type	magnetophonon resonance	85S
	271	LO, GaAs-type		
electron mobility (in cm^2/Vs):				
typical values from papers on characterization of layers:				
μ_n	13 800	300 K	$n = 2\cdot10^{15}\,cm^{-3}$, LPE-layer	81O
	70 000	77 K		
	11 200	300 K	$n \approx 10^{15}\,cm^{-3}$, MOCVD-layer	85C
	64 000	77 K		

Gallium indium antimonide ($Ga_xIn_{1-x}Sb$)

The properties of this system are in many respect similar to those of $Ga_xIn_{1-x}As$. Both components are direct gap semiconductors.

energy gap (in eV):				
$E_g(x, 0)$	$0.235 + 0.1653\,x$ $+ 0.413\,x^2$	0 K	photovoltaic effect	80R

Physical property	Numerical value	Experimental conditions	Experimental method, remarks	Ref.

temperature dependence:

$$E_g(x, T) = E_g(x, 0) + a_1 T^2 x/(b_1 + T) + a_2 T^2(1-x)/(b_2 + T) - x(1-x)c_T T$$

a_1	$-6.2(1) \cdot 10^{-4}\,\mathrm{eV\,K^{-1}}$		80R
a_2	$-3.8(1) \cdot 10^{-4}\,\mathrm{eV\,K^{-1}}$		
b_1	$260(20)\,\mathrm{K}$		
b_2	$200(20)\,\mathrm{K}$		
c_T	$-5(2) \cdot 10^{-5}\,\mathrm{eV\,K^{-1}}$		

References for 2.16.1

70L1   Lucovski, G., Chen, M.F.: Solid State Commun. **8** (1970) 1397.
70L2   Lorenz, M.R., Onton, A.: Proc. 10th Int. Conf. Semicond. Phys., Cambridge/Mass., USAEC, New York 1970, p. 444.
70O   Onton, A., Chicotka, R.J.: J. Appl. Phys. **41** (1970) 4305.
78O   Olsen, G.H., Nuese, C.J., Smith, R.T.: J. Appl. Phys. **49** (1978) 5523.
79K   Kano, H., Miyazawa, S., Sugiyama, K.: Jpn. J. Appl. Phys. **18** (1979) 2183.
80A   Alavi, K., Aggarwal, R.L., Groves, S.H.: Phys. Rev. **B21** (1980) 1311.
80R   Roth, A.P., Keeler, W.J., Fortin, E.: Can. J. Phys. **58** (1980) 560.
81O   Oliver, J.D.: J. Cryst. Growth **54** (1981) 64.
81R   Roberts, J.S., Scott, G.B., Gowers, J.P.: J. Appl. Phys. **52** (1981) 4018.
81S   Saxena, A.K.: Phys. Status Solidi (b) **105** (1981) 777.
83A   Alibert, C., Joullié, A., Joullié, A.M., Ance, C.: Phys. Rev. **B27** (1983) 4946.
83B1   Biryulin, Yu.F., Ganina, N.V., Mil'vidskii, M.G., Chaldychev, V.V., Shmartsev, Yu.V.: Sov. Phys. Semicond. (English Transl.) **17** (1983) 68; Fiz. Tekh. Poluprovodn. **17** (1983) 108.
83B2   Biryulin, Yu.F., Vul', S.P., Chaldychev, V.V., Shmartsev, Yu.V.: Sov. Phys. Semicond. (English Transl.) **17** (1983) 65; Fiz. Tekh. Poluprovodn. **17** (1983) 103.
83B3   Bugajski, M., Kontkiewicz, A.M., Mariette, H.: Phys. Rev. **B28** (1983) 7105.
83G   Goetz, K.-H., Bimberg, D., Jürgensen, H., Selders, J., Solomonov, A.V., Glinskii, G.F., Razhegi, M.: J. Appl. Phys. **54** (1983) 4543.
83P   Pearsall, T.P., Carles, R., Portal, J.C.: Appl. Phys. Lett. **42** (1983) 436.
84D   Davies, G.J., Kerr, T., Tuppen, C.G., Wakefield, B., Andrews, D.A.: J. Vac. Sci. Technol. **B2** (1984) 219.
84W   Wakefield, B., Halliwell, M.A.G., Kerr, T., Andrews, D.A., Davies, G.J., Wood, D.R.: Appl. Phys. Lett. **44** (1984) 341.
84Y   Yu, P.W., Kuphal, E.: Solid State Commun. **49** (1984) 907.
85A   Adachi, S.: J. Appl. Phys. **58** (1985) R1.
85C   Chan, K.T., Zhu, L.D., Ballantyne, J.M.: Appl. Phys. Lett. **47** (1985) 44.
85K   Kuo, C.P., Vong, S.K., Cohen, R.M., Stringfellow, G.B.: J. Appl. Phys. **57** (1985) 5428.
85S   Sarkar, C.K., Nicholas, R.J., Portal, J.C., Razhegi, M., Chevrier, J., Massies, J.: J. Phys. **C18** (1985) 2667.
85Z   Zarrabi, H.J., Alfano, R.R.: Phys. Rev. **B32** (1985) 3947.

2.16.2 Ternary alloys of the type III–V_{1-x}–V_x

Aluminum arsenide phosphide (AlAs$_{1-x}$P$_x$)

The only interest in this system up to now has been an improvement of the lattice match of AlAs on GaAs by adding a small amount of phosphorus. AlAs$_{0.96}$P$_{0.04}$ layers made by MOCVD lattice-match exactly on GaAs at RT (Fig. 1) [84K].

Gallium arsenide phosphide (GaAs$_{1-x}$P$_x$)

Alloys of this type are of technological interest when properties of GaAs should be combined with a higher direct energy gap material. Above $x = 0.45$ GaAs$_{1-x}$P$_x$ changes to an indirect gap semiconductor.

energy gaps (in eV, relative to the top of valence band):

$E_{gx}(\Gamma)$	$1.515 + 1.172\,x$ $+ 0.186\,x^2$	2 K	excitonic gap from photo-luminescence	81C
$E_{gx}(X)$	$1.9715 + 0.144\,x$ $+ 0.211\,x^2$			

Physical property	Numerical value	Experimental conditions	Experimental method, remarks	Ref.

Gallium arsenide antimonide ($GaAs_{1-x}Sb_x$)

Both components are direct gap semiconductors. The system has a miscibility gap with a peritectic temperature of 751 °C [84C] limiting the range of composition which can be obtained by near equilibrium growth techniques at ordinary temperatures.

energy gap (in eV):

E_g	$1.43 - 1.9\,x + 1.2\,x^2$	RT	photoluminescence	77N

effective mass, electrons (in units of m_0):

m_n	$0.0634 - 0.0483\,x$ $+ 0.0252\,x^2$		$x \leqq 0.6$, Faraday rotation, plasma resonance	81D

Indium arsenide phosphide ($InAs_{1-x}P_x$)

This system has received considerable attention since the energy gaps cover the infrared spectrum. Both components are direct gap semiconductors.

energy gap (in eV):

E_g	$0.356 + 0.675\,x$ $+ 0.32\,x^2$	300 K	transmission	79N
	$0.414 + 0.64\,x + 0.36\,x^2$	77 K		

electron mobility:

Electron mobilities in epitaxial layers up to $3 \cdot 10^5$ cm^2/Vs at 77 K for $x = 0.87$ have been reported [84W].

Indium antimonide arsenide ($InSb_{1-x}As_x$)

Only a few papers on this system have been published.

Indium antimonide bismuthide ($InSb_{1-x}Bi_x$)

Since InBi is semimetallic with a tetragonal structure only a limited range of solid solutions show semiconducting behavior ($x < 0.03$). Metastable films have been investigated

References for 2.16.2

77N   Nahory, R.L., Pollack, M.A., DeWinter, J.C., Williams, K.M.: J. Appl. Phys. **48** (1977) 513.
79N   Nicholas, R.J., Stradling, R.A., Ramage, J.C.: J. Phys. **C12** (1979) 1641.
81C   Capizzi, M., Modesti, S., Martelli, F., Frova, A.: Solid State Commun. **39** (1981) 333.
81D   Devlin, P., Heravi, H.M., Woolley, J.C.: Can. J. Phys. **59** (1981) 939.
84K   Kobayashi, N., Fukui, T.: J. Cryst. Growth **67** (1984) 513.
84C   Cherng, M.J., Stringfellow, G.B., Cohen, R.M.: Appl. Phys. Lett. **44** (1984) 677.
84W   Wang, P.J., Wessels, B.W.: Appl. Phys. Lett. **44** (1984) 766.

2.16.3 Quaternary alloys of the type $III_x-III_{1-x}-V_y-V_{1-y}$

Quaternary alloys provide the possibility to grow epitaxial layers with a broad range of energy gaps lattice matched to a suitable substrate.

The condition for lattice matching can easily be derived from the interpolation scheme for the determination of a material parameter $P(x, y)$ for an alloy $A_x B_{1-x} C_y D_{1-y}$ from the same parameters of the four constituents:

$$P(x, y) = (1 - x)y\,P(BC) + (1 - x)(1 - y)P(BD) + xy\,P(AC) + x(1 - y)P(AD).$$

Physical property	Numerical value	Experimental conditions	Experimental method, remarks	Ref.

For lattice matching on substrate BD (x = y = 0) the condition $a(x, y) = a(BD)$ leads to

$$x = \frac{[a(BC) - a(BD)]y}{[a(BD) - a(AD)] - [a(BC) + a(AD) - a(BD) - a(AC)]y}$$

or, in a linear approximation to this formula – exact at y = 0 and y = 1:

$$x/y = [a(BC) - a(BD)]/[a(BC) - a(AC)].$$

The main interest in quaternary III–V alloys stems from the possible applications in micro- and optoelectronic devices. Thus most of the data in the following sections refer to quaternary alloys lattice matched to GaSb, InP and GaAs (see Fig. 1 in section 2.16.1).

Several types of quaternary alloys are possible:

(a) III-III-V-V alloys. These materials will be discussed in the following subsections.

(b) III-III-III-V alloys. Several systems will be presented in section 2.16.4.

(c) III-V-V-V alloys. Here only one system seems to be of interest for applications: $InAs_{1-x-y}Sb_xP_y$.

Not much has been done in the field of *quintarnary alloys*. An example for such material is $(Al_xGa_{1-x})_{1-z}In_zP_yAs_{1-y}$ on GaAs [84M].

Gallium indium arsenide phosphide ($Ga_xIn_{1-x}As_yP_{1-y}$)

As shown in Fig. 1 of section 2.16.1 alloys of this system can be lattice matched on InP (E_g range: $0.73\cdots1.35\,eV$), GaAs (E_g range: $1.42\cdots1.90\,eV$) and ZnSe (slightly lower range). The matching conditions according to the formula above are ($0 \leq y \leq 1$):

x	0.1894 y/ (0.4184 − 0.013 y) ≈ 0.47 y		on InP substrate
	(1.00 + y)/2.08		on GaAs substrate
	(1.06 + y)/2.06		on ZnSe substrate

Most data have been obtained for InP lattice matched samples. All data in the following tables refer – if not stated otherwise – to InP lattice matched material.

direct energy gap (in eV):

$E_{g,dir}(x, y)$	$1.35 + 0.668\,x - 1.068\,y + 0.758\,x^2 + 0.078\,y^2 - 0.069\,xy - 0.322\,x^2y + 0.03\,xy^2$			84K1

from literature data for the constituents and interpolation formulas; for lattice match to InP this formula reduces to

$E_{g,dir}(y)$	$1.35 - 0.775\,y + 0.149\,y^2$	298 K	calculated, fitting photoluminescence and electroreflectance measurements,	82P
	$1.425 - 0.7668\,y + 0.149\,y^2$	4.2 K	calculated, fitting absorption and transmission measurements.	

bowing parameters for band gaps at L and X (in eV):

$c(\Gamma - L)$	0.10(5)		L-conduction band	84K2
$c(\Gamma - X)$	0.21(7)		X-conduction band from synchrotron radiation reflection spectroscopy	

higher interband transition energies (in eV):

E_1	$3.11 + 0.034\,x - 0.885\,y + 0.516\,x^2 + 0.275\,y^2 - 0.187\,xy + 0.017\,x^2y$		80L

Physical property	Numerical value	Experimental conditions	Experimental method, remarks	Ref.

at RT, from fitting of literature data: for InP lattice matched material this formula reduces to

| $E_1(y)$ | $3.11 - 0.87\,y + 0.30\,y^2$ $+ 0.007\,y^3$ | RT | | 80L |

spin-orbit splitting energies (in eV):

| $\Delta_0(y)$ | $0.119 + 0.300\,y$ $- 0.107\,y^2$ | RT | electroreflectance, | 80P |
| $\Delta_1(y)$ | $0.145 + 0.173\,y$ $- 0.064\,y^2$ | | | 80P |

effective masses (in units of m_0):

| $m_n(x, y)$ | $0.08 - 0.116\,x + 0.026\,y - 0.059\,xy + (0.064 - 0.02\,x)y^2 + (0.06 + 0.032\,y)x^2$ | | | 80R |

interpolation formula; a relation for InP lattice matched material is

| $m_n(y)$ | $0.077 - 0.050\,y$ $+ 0.014\,y^2$ | RT | Shubnikov – de Haas effect | 80P |

lattice parameter (in Å):

| $a(x, y)$ | $5.8688 - 0.4176x$ $+ 0.1896y + 0.0125y$ | | linear interpolation from lattice parameters of four constituents | 82A1 |

density (in g cm^{-3}):

| d | $5.477 - 0.712y$ | | | 82A1 |

dielectric constant:

| $\varepsilon(0)$ | $12.40 + 1.5y$ | | | 82A2 |
| $\varepsilon(\infty)$ | $9.55 + 2.2y$ | | | |

Gallium indium arsenide antimonide ($Ga_x In_{1-x} As_y Sb_{1-y}$)

This system is shown as shaded area in Fig. 1 of section 2.16.1. It is the only system applicable for the growth of low band gap epitaxial layers of GaSb substrate. In spite of interesting applications for optical sources and detectors in the $2 \cdots 4\,\mu m$ range only few reliable data on intrinsic properties have been published.

Aluminum gallium arsenide antimonide ($Al_x Ga_{1-x} As_y Sb_{1-y}$)

This system shows a broad miscibility gap. Epitaxial layer growth is possible
– *on GaSb* for small amounts of As; by adding small amounts of As the lattice matching of $Al_x Ga_{1-x}Sb$ on GaSb can be improved (see Fig. 1 of section 2.16.1)
– *on GaAs* for small amounts of Sb; layers with y greater than 0.8 have been produced.

References for 2.16.3

80L   Laufer, P.M., Pollack, M.A., Nahory, R.E.: Solid State Commun. **36** (1980) 419.
80P   Perea, E.H., Mendez, E.E., Fonstad, C.G.: Appl. Phys. Lett. **36** (1980) 978.
80R   Restorff, J.B., Houston, B., Allgaier, R.S.: J. Appl. Phys. **51** (1980) 2277.
82A1   Adachi, S.: J. Appl. Phys. **53** (1982) 8775.
82A2   Adachi, S.: J. Appl. Phys. **53** (1982) 5863.
82P   Pearsall, T.P.: "GaInAsP Alloy Semiconductors", T.P. Pearsall ed., J. Wiley & Sons, New York **1982**, p. 295.
84K1   Kuphal, E.: J. Cryst. Growth **67** (1984) 441.
84K2   Kelso, S.M., Aspnes, D.E., Olson, C.G., Lynch, D.W., Bachmann, K.J.: Proc. SPIE Int. Soc. Opt. Eng. **452** (1984) 200.
84M   Mukai, S., Yajima, H., Mitsuhashi, Y., Yanagisawa, S., Kutsuwada, N.: Appl. Phys. Lett. **44** (1984) 904.

Physical property	Numerical value	Experimental conditions	Experimental method, remarks	Ref.

2.16.4 Quaternary alloys of the type $III_{1-x-y}–III_x–III_y–V$

Indium aluminum gallium phosphide ($In_{1-x-y}Al_xGa_yP$)

The boundary alloys of the GaAs lattice matched series are $In_{0.49}Ga_{0.51}P$ and $In_{0.49}Al_{0.51}P$ (see Fig. 1 of section 2.16.1). Introducing the compositional parameter z by $x = 0.51\,z$ and the lattice matching condition $x + y = 0.51$ the series can be written as

$$In_{0.49}(Al_zGa_{1-z})_{0.51}P.$$

energy gap (in eV):

E_g	$1.9 + 0.6z$	RT		82A

electron effective mass (in units of m_0):

m_n	$0.0427\,(15)$ $+0.0328\,(7)\,z$			82O1, 82O2

Indium aluminum gallium arsenide ($In_{1-x-y}Al_xGa_yAs$)

The boundary alloys of the InP lattice matched series are $In_{0.53}Ga_{0.47}As$ and $In_{0.52}Al_{0.48}As$ (see Fig. 1 of section 2.16.1). Introducing the compositional parameter z by $x = 0.48\,z$ and the lattice matching condition $0.98x + y = 0.47$ the series can be written as

$$(In_{0.52}Al_{0.48})_z(In_{0.53}Ga_{0.47})_{1-z}As.$$

An often used approximation is

$$In_{0.53}(Al_zGa_{1-z})_{0.47}As \quad \text{with } x = 0.47z, \quad x + y = 0.47$$

energy gap (in eV):

$E_g(x, y)$	$0.36 + 2.093x + 0.629y + 0.577x^2 + 0.436y^2 + 1.013xy - 2.0xy(1 - x - y)$		82O1

For the lattice matching condition (given in [82O1] more exactly as $0.983x + y = 0.468$) using $z = x/0.48$ the experimental data obtained from photoluminescence at RT give:

$E_g(z)$	$0.76\,(4) + 0.49\,(5)z + 0.20\,(3)z^2$		82O1, 82O2

Indium aluminum gallium antimonide ($In_{1-x-y}Al_xGa_ySb$)

For material prepared by directional freezing and investigated by X-ray and optical methods the following relations have been obtained ($z = 1 - x - y$):

a[nm]	$0.64789 - 0.03433x - 0.03834y - xyz(0.025 + 0.325z)$		81Z
$E_{g,dir}$	$0.095 + 1.76x + 0.28y + 0.345(x^2 + y^2) + 0.085z^2 + xyz(23 - 28y)$		
$E_{g,ind}$	$1.0675 + 0.30x - 0.31y + 0.2625(x^2 + y^2 + z^2) + xyz(20x - 5.9)$		

References for 2.16.4

81Z   Zbitnew, K., Woolley, J.C.: J. Appl. Phys. **52** (1981) 6611.
82A   Asahi, H., Kawamura, Y., Nagai, H.: J. Appl. Phys. **53** (1982) 4928.
82O1   Olego, D., Chang, T.Y., Silberg, E., Caridi, E.A., Pinczuk, A.: Int. Phys. Conf. Ser. No. 65, Int. Symp. GaAs and Related Compounds, Albuquerque **1982**, p. 195.
82O2   Olego, D., Chang, T.Y., Silberg, E., Caridi, E.A., Pinczuk, A.: Appl. Phys. Lett. **41** (1982) 476.

Appendix: Contents of the volumes of the New Series of Landolt-Börnstein dealing with group IV and III-V semiconductors

In the following volumes of the New Series of Landolt-Börnstein extensive material is presented about the properties of semiconducting group IV elements and III-V compounds.

Volume III/17a: Physics of Group IV Elements and III-V Compounds (642 pages, published 1982)

Editor: O. Madelung
Authors: D. Bimberg, R. Blachnik, M. Cardona, P.J. Dean, Th. Grave, G. Harbeke, K. Hübner, U. Kaufmann, W. Kress, O. Madelung, W. von Münch, U. Rössler, J. Schneider, M. Schulz, M.S. Skolnick

Volume III/22a: Intrinsic Properties of Group IV Elements and III-V, II-VI and I-VII Compounds, Supplement and Extension to Volumes III/17a and b (452 pages, published 1987)

Editor: O. Madelung
Authors: O. Madelung, W. von der Osten, U. Rössler

Volume III/22b: Impurities and Defects in Group IV Elements and III-V Compounds, Supplement and Extension to Volumes III/17a, c and d (776 pages, published 1989)

Editor: M. Schulz
Authors: C.A.J. Ammerlaan, W. Bergholz, B. Clerjaud, H. Ennen, H.G. Grimmeis, B. Hamilton, U. Kaufmann, W. von Münch, R. Murray, R.C. Newman, A.R. Peaker, G. Pensl, H.-J. Rath, R. Sauer, J. Schneider, M. Schulz, M.S. Skolnick, N.A. Stolwijk, P. Vogl, A.F.W. Willoughby, W. Zulehner

Volume III/23a: Electronic Structure of Solids: Photoemission Spectra and Related Data (430 pages, published 1989)

Editor: A. Goldmann
Authors of Chapter 2.1: T.C. Chiang, F.J. Himpsel

Furthermore the **Volumes III/17c (Technology of Si, Ge and SiC) and III/17d (Technology of III-V, II-VI and Non-Tetrahedrally Bonded Semiconductors)** contain data on group IV and III-V semiconductors.

We present in the following a short review about the organization and contents of volumes III/17a, III/22a, III/22b and III/23a:

Volume III/17a: Physics of Group IV Elements and III-V Compounds

Organization of the volume:
The semiconductors are grouped into families according to the chemical nature of the constituents: IVth group elements, III-V compounds, II-VI compounds etc.
   Within each chapter the organization is as follows:

"0-section": Presentation of all members of the group, review of general relationships, structure, non-semiconducting phases (or members of the family not being semiconductors), high-pressure phases etc. General discussion of chemical bond within the group.
The 0-section is followed by separate sections on each member of the group and on solid solutions within the group and with members of other groups. The physical properties are listed – as far as possible – in subsections of the following order:

1 Electronic properties: Information and data about electronic and excitonic energy states as well as electron and hole parameters:

band structure, density of states / band gaps / exciton data / spin-orbit splitting energies / intraband and interband transition energies / effective masses and g-factors of electrons / effective masses of holes, other valence band parameters / deformation potentials.

2 Impurities and defects: Basic data on shallow and deep states, bound excitons and local modes (data on diffusion and distribution coefficients are presented in the technological chapters of subvolume 17c):

shallow donors (ionization energies, excited states, ground state splittings, deformation potentials etc.) / bound excitons (localization energies, capture cross-sections, transition lifetimes etc.) / local modes (energies, isotope shifts etc.) / deep traps, 3d transition metals (ionization energies, esr and endor data etc.)

3 Lattice properties: Static and dynamical properties of the lattice (for structure, phase transition, chemical bond, see the 0-section of the respective chapter; for density and melting point, see subsection 6; for static dielectric constant, see subsection 5):

lattice parameter, thermal expansion / phonon dispersion relations, density of states / phonon frequencies (wavenumbers) / sound velocities / elastic moduli (also third order) / Young's modulus, torison modulus, bulk modulus, compressibility, Poisson's ratio, Grüneisen constant, mode Grüneisen parameters, effective ion charges etc.

4 Transport properties: Electronic transport parameters (for thermal conductivity, see subsection 6): conductivity, carrier concentration / electron and hole mobilities, warm electron coefficient / Hall effect and magnetoresistance / piezo- and elastoresistance, elastostriction, piezoelectric coefficients / other transport parameters as Seebeck coefficient (thermoelectric power), Nernst coefficient etc.

5 Optical properties: Optical spectra, optical constants, parameters obtained from optical experiments (if not already presented in subsections 1 and 2):

refractive index, absorption index / absorption coefficient and reflectance / dielectric constant (including static dielectric constant) / optical spectra / other optical coefficients as Verdet coefficient, two-photon absorption coefficient, piezooptic coefficients etc. / Raman and Brillouin scattering / electron energy loss / optical spectra including core levels (vacuum uv spectra, yield spectry, ESCA, XPS, UPS, Auger spectra etc.) / Schottky barrier heights.

6 Further properties: Thermal, magnetic, thermodynamic properties, some other general data: thermal conductivity / magnetic susceptibility / Debye temperature, heat capacity / hardness / density / melting point / thermodynamical data (characteristical data on fusion, vaporization, entropy etc.)

7 References

Volume III/22a: Intrinsic Properties of Group IV Elements and III-V, II-VI and I-VII Compounds

This volume contains in its first two chapters data on the intrinsic properties of group IV elements and III-V compounds published within the years 1980 to 1986. Some earlier data have been included if it seemed necessary for the convenience of the user.

The table of contents of this volume is identical with the organization scheme of volume III/17a. Each section is organized in the following subsections:

1 Electronic properties / 2 Lattice properties / 3 Transport properties / 4 Optical properties / 5 References.

This scheme is different from the organization scheme in volume III/17a by two items: (a) Subsections on impurities and defects are lacking since such items are covered by subvolume III/22b. (b) The section "Structure and chemical bond" and the subsection "Further properties" have been cancelled. Information about these topics is now included in the subsections "Lattice properties" (structure, phases, melting point, density, thermochemical properties etc.) and "Transport properties" (thermal conductivity).

Table of Contents of volume III/17a and the first two chapters of volume III/22a

Volume III/22b: Impurities and Defects in Group IV Elements and III-V Compounds

This data collection contains critically reviewed information on impurities and defects in elemental and III-V compound semiconductors. Electrical, optical, and chemical methods for the characterization of defects are briefly outlined. Comprehensive references to the literature and to examples for applications are given. Trends of defect properties in similar materials and chemical trends for the properties of common defects are outlined in a theoretical chapter.

Data are only presented for technically important elemental semiconductors in group IV of the periodic table and of the III-V compounds.

Chapters 2 surveys the trends of impurity and defect properties as predicted by theory. A brief review of theoretical methods for defect level calculation is given. Chapter 3 presents an overview on the measurement and analysis of defect properties. Electrical, optical, and chemical methods are briefly introduced and surveyed by their properties for defect characterization. Defect and impurity data are listed in the comprehensive sections 4···7.

For each material the properties are listed roughly in the following order:
1 Solubility and segregation of impurities / 2 Diffusion of impurities / 3 Impurity levels / 4 Excited bound states of acceptors and donors / 5 Photoluminescence properties of impurities and defects / 6 Paramagnetic centers / 7 Vibrational modes of impurities and defects / 8 Oxygen-related defects and microdefects.

Table of Contents of Volume III/22b

Volume III/23a: Electronic Structure of Solids: Photoemission Spectra and Related Data

Organization of the volume:
Within each chapter the organization is as follows: First, general data (as far as available) are summarized on crystal structure, electronic configuration, work functions, plasmon energies, core level binding energies, valence band critical point energies, and on other relevant quantities. Then diagrams are collected reproducing angle-integrated as well as angle-resolved valence band and core level spectra, calculated energy bands and corresponding densities of states, and in particular experimental electron energy dispersion curves $E(k)$. When considered necessary, also optical spectra and results obtained with other experimental techniques are shown to supplement the electronic structure information.

Table of Content of Chapter 2.1 of Volume III/23a